A Place in Space

A Place in Space

Ethics, Aesthetics,
and Watersheds

New and Selected Prose

Gary Snyder

COUNTERPOINT

WASHINGTON, D.C.

The author wishes to thank and acknowledge
the editors and publishers
who first published many of these essays
in earlier drafts and incarnations:
American Poetry, *Audubon*, Casa de América,
City Lights Books, Empty Bowl Press,
Library Perspective, *Mesechabe*,
New Directions Publishing, North Point Press,
San Francisco Examiner, Shambhala Publications,
Sulfur, *Ten Directions*, *Tree Rings*,
Upriver Downriver, Weatherhill, Inc., and *ZYZZYVA*.

Library of Congress Cataloging-in-Publication Data
Snyder, Gary, 1930–
A place in space: ethics, aesthetics, and watersheds:
new and selected prose / Gary Snyder.
I. Title.
PS3569.N88P57   1995
814´.54—dc20   95-32305
ISBN 1-887178-27-9 (alk. paper)

Book design by David Bullen
Composition by Wilsted & Taylor
Printed in the United States of America on acid-free paper that
meets the American National Standards Institute Z39-48 Standard.

COUNTERPOINT
P.O. Box 65793
Washington, D.C. 20035-5793

Distributed by Publishers Group West

Contents

Note

*T*his collection draws on some forty years of thinking and writing. It can be considered a further exploration of what the "practice of the wild" would be.

The ancient Buddhist precept "Cause the least possible harm" and the implicit ecological call to "Let nature flourish" join in a reverence for human life and then go beyond that to include the rest of creation. These essays are Buddhist, poetic, and environmental calls to complex moral thought and action—metaphoric, oblique, and mythopoetic, but also I hope practical. Ethics and aesthetics are deeply intertwined. Art, beauty, and craft have always drawn on the self-organizing "wild" side of language and mind. Human ideas of place and space, our contemporary focus on watersheds, become both model and metaphor. Our hope would be to see the interacting realms, learn where we are, and thereby move toward a style of planetary and ecological cosmopolitanism.

Meanwhile, be lean, compassionate, and virtuously ferocious, living in the self-disciplined elegance of "wild mind."

I

Ethics

North Beach

*I*n the spiritual and political loneliness of America of the fifties you'd hitch a thousand miles to meet a friend. Whatever lives needs a habitat, a culture of warmth and moisture to grow. West Coast of those days, San Francisco was the only city; and of San Francisco, our home port was North Beach. Why? Because partly, totally non-Anglo. First, the Costanoan native peoples—peoples living around the Bay for five thousand–plus years. Sergeant José Ortega crossed sand dunes and thickets to climb a hill (Telegraph) there around the first of November, 1769. Later, Irish on the hill (prior to quake and fire) and tales of goats grazing those rocks—

> Tellygraft hill, Tellygraft hill,
> Knobby old, slobby old,
> Tellygraft hill

—then Italian, Sicilian, Portuguese (fishermen), Chinese (Kuang-tung and Hakka), and even Basque sheepherders down from Nevada on vacation.

When we of the fifties and after walked into it, *walk* was

the key word. Maybe no place else in urban America has such a feel of on-foot: narrow streets, high blank walls, and stair-step steeps of alleys and white-wood houses cheap to rent; laundry flapping in the foggy wind from flat-topped roofs. Like Morocco, or ancient terraced fertile-crescent pueblos.

A tiny watershed divide is at the corner of Green and Columbus. Northward a creek flowed, the mouth of which, on the little alley called Water Street (now some blocks up from the Fisherman's Wharf coast—all fill), is under the basement of a friend's apartment. The easterly stream went down by what was the Barbary Coast strip of clubs once, and on underneath the present Coast and Geodetic Survey offices on Battery. Storms come out of a place in the North Pacific, high latitudes, pulse after pulse of weather (storms deflected north in summer). San Francisco, North Beach, like living on the bow of a ship. Over the dark running seas, from November on, breaking in rains and flying cloud bits on the sharp edges of Telegraph Hill.

A habitat, midway between two other summer and winter ranges, Berkeley and Marin County. Who would not, en route, stop off in North Beach? To buy duck eggs, drop into Vesuvio's, City Lights, get sesame oil or wine, walk up Grant to this or that Place. Or living there: the hum of cable-car cables under the street—lit-up ships down on the docks working all night—the predawn crashes of the Scavengers' trucks. Spanning years from a time when young women would get arrested for walking barefoot, to the barebottom clubs of Broadway now tending tourist tastes from afar.

A habitat. The Transamerica Pyramid, a strikingly wasteful and arrogant building, stands square on what was once called Montgomery Block, a building that housed the artists and revolutionaries of the thirties and forties. Kenneth Rexroth, many others, lived there; foundations of postwar libertarianism; movements that became publicly known as "beat" in the middle fifties. This emphasis often neglected the deeply dug-in and committed thinkers and artists of the era who never got or needed much media fame; who were the *culture* that nourished so much. Many people risking all—following sometimes the path of excess and not always going beyond folly to the hoped-for wisdom. Yet, like the sub-Aleutian storms, pulse after pulse came out of North Beach from the fifties forward that touched the lives of people around the world.

I worked the docks in those days.

Down to Pier 23 to work. Smith-Rice cranes, and Friday a white egret that fluttered down on the pier, dwarfing the seagulls, riffled its wings and feathers delicately a few times then flew off back in the direction from which it came.—23.XI.52

It is of no particular significance that I sit writing Chinese characters and practice pronouncing them in Japanese; it's all here: vines in the Mediterranean, taro patches in Melanesia, the clover yards of Vancouver Island—the eye sees, the hand moves, the world moves in and through, like a complex spiral shell.—4.II.54

And a *People's World* headline from October 1958:

> Outmoded Capitalism
> Threatens Humanity
> With Multiple Perils

while walking to Gino and Carlo's, another place we met and drank (Jack Spicer gave me a whiskery hug).

> *The necessity to roam at wild . . . large, useless, and no-*
> *where scenes, to follow the city cat-track down, "out of my*
> *head," etc.—we need the big gamble of a physical eco-*
> *nomic urban Void in which you have to dive.—3.XI.58*

That close, loose circle of comrades, lovers, freaks, and friends (how many we mourn already!) in the rolling terrain of North Beach—virtually the only place in California that didn't freeze out plants in the cold snap of December 1972; in fact, warmer than anyplace else in San Francisco except for Noe Valley, and having the most frost-free days per year of any place in the U.S. short of Florida—is the rich soil of much beauty, and the good work of hatching something *else* in America; pray it cracks the shell in time. Gratitude to the Spirits of the Place; may all Beings flourish.

[This was written for an ingenious event called "North Beach as Hab-itat," organized by architects Dan Osborne and Zach Stewart and held at the Canessa Gallery on Montgomery Street in January 1975. The es-say was then published in The Old Ways *(San Francisco: City Lights, 1977).]*

"Notes on the Beat Generation" and "The New Wind"

NOTES ON THE BEAT GENERATION

One day in September 1955, I was fixing my bicycle in the yard of a little cabin in Berkeley, California—just back from three months working on a trail crew in the Sierra Nevada—and a respectable-looking fellow in a dark business suit came around the corner and asked if I was Gary Snyder and said his name was Allen Ginsberg. It was the only time I ever saw Allen Ginsberg in a suit. We had tea together, and he said he had come from the poet Kenneth Rexroth to see me, with the idea of getting a few San Francisco poets together to hold a poetry reading in a little art gallery in the City. Two weeks later Phil Whalen came hitchhiking into town from the mountains of Washington, and at the same time Jack Kerouac arrived, riding a freight train up from Los Angeles. Ginsberg had moved to Berkeley (the site of the University of California, and only half an hour by bus from San Francisco) to enter the graduate Department of English, with the idea of settling down and having a career as a professor. He stayed with it about three weeks and quit. It was no time for any of us to worry about college degrees or what kind of work we would do in the future; it was time to write poetry. In the month of October 1955, with wine, marijuana, and jazz, Ginsberg wrote his now-famous poem *Howl*, and at the end of October we gave our poetry reading. Philip Lamantia (whose new book of poems *Ekstasis* is just out), Mike McClure, Whalen, Ginsberg, and I read. Kerouac was there beating tunes on empty bottles, and Rexroth was cracking jokes about his diplomat's suit that he'd bought secondhand for a few dollars. All of us had five to ten years' worth of poems on hand, and most of them had never been heard by anyone. That night of poetry, with the wild crowd that came, was the beginning of the "San Francisco poetry renaissance"—which has since become part of the whole phenomenon known as the "beat generation." From that night on, there was a poetry reading in somebody's pad, or some bar or gallery, every week in San Francisco. We had a sudden feeling that we had finally broken through to a new freedom of expression, had shattered the stranglehold of universities on poets, and gone beyond the tedious and pointless arguments of Bolshevik versus capitalist that were (and still are) draining the imaginative life out of so many intellectuals in the world. What we had discovered, or rediscovered, was that the imagination has a free and spontaneous life of its own, that it can be trusted, that what flows from a spontaneous mind is poetry—and that this is more basic and more revolutionary than any political program based on "civilized abstractions" that end up murdering human beings in the name of historical necessity or Reason or Liberty; Russia and America are both huge witless killers of the heart of man.

Jack Kerouac, at that time, was living the hobo life—one pair of blue jeans, a pack on his back, and a notebook to write in. But even then he had the manuscripts of ten or twelve completed novels stored away. When his novel *On the Road* was published in 1957, the word *beat* became famous and overnight America became aware that it had a generation of writers and intellectuals on its hands that was breaking all the rules. This new generation was educated, but it refused to go into academic careers or business or government. It published its poems in its own little magazines, and didn't even bother to submit works to the large established highbrow journals that had held the monopoly on avant-garde writing for so long. Its members traveled easily, bumming from New York to Mexico City to San Francisco—the big triangle—and traveled light. They stayed with friends in San Francisco's North Beach or on New York's Lower East Side (the Greenwich Village of the beat, really a slum)—and made their money at almost any kind of work. Carpentry, railroad jobs, logging, farm work, dishwashing, freight handling—anything would do. A regular job ties you down and leaves you no time. Better to live simply, be poor, and have the time to wander and write and *dig* (meaning to penetrate and absorb and enjoy) what was going on in the world. *On the Road* (translated into Japanese as *Rojo*) describes much of this life. The people who read it were either scared or delighted. Many of the delighted ones moved out to San Francisco (scene of Kerouac's subsequent novel, *The Subterraneans*, which has been translated into Japanese as *Chikagai no Hitobito*) to join in the fun. As for Ginsberg's *Howl*, after initially being banned in San Francisco by the police, it won a court trial and became a poetry best-seller among the younger generation, who saw in it a poem that spoke directly to them. "Beat generation" became a household term.

What was the reaction of the newspapers and the public to this? They were either outraged or a little jealous. It is one of those few times in American history that a section of the population has freely chosen to disaffiliate itself from "the American standard of living" and all that goes with it—in the name of freedom. And in a way, the literature of the beat generation is some of the only true proletarian literature in recent history—because actual members of the working class are writing it, "proletarian bohemians" if you will; and it is not the sort of thing that middle-class Communist intellectuals think proletarian literature ought to be. There is no self-pity or accusation or politics, simply human beings and facts. The class struggle means little to those who have abandoned all classes in their own minds and lives. But both the left-wing intellectuals and the newspaper editors of America see this as heresy of the worst sort, and both sides shout "irresponsible." They point to the jazz world as decadent (whereas it is in fact one of the most creative and profound things in America) and write about the sexual immorality and delinquency and use of drugs by the young writers of

San Francisco and New York. To such charges, the answers are simple: these people are interested in revolution, *real* revolution, which starts with the individual mind and body. One of the things that has been dragging the soul of the world down since the end of the Bronze Age is the family system and associated notions of sexual morality that go with patrilineal descent and the descent of property in the male line (anybody who has read Engels's *Origins of the Family, Private Property, and the State* knows what I am talking about). There will be no economic revolution in this world that works, without a sexual revolution to go with it. Any person who attempts to discover in practice what the real values of sex are, and what marriage really means, will be called (as D. H. Lawrence was) immoral or obscene. As for delinquency, or crime, this is largely slander. Juvenile delinquents, as Lawrence Lipton points out in his book *The Holy Barbarians*, are simply teenage capitalists who want their money right away. The bohemian men and women of San Francisco's North Beach buy their clothes in secondhand stores, and none of them own radios or cars. The last charge, drugs, is a difficult point, because marijuana is classified as a dangerous narcotic in America, and hence is illegal. Marijuana (Indian hemp, *Cannabis sativa*), along with cheap red wine, is a standby of beat social life. It is either imported from Mexico or grown at home, and the penalties for possessing are very severe (at least six months in prison). But everybody smokes it—calls it "pot" or "tea" or "grass"—to enjoy the quiet sharpening of perceptions it gives, particularly to the ear, which explains why marijuana is so widely used by jazz musicians. The public is not informed of the fact that marijuana has been pronounced by doctors as non–habit forming, non–tolerance forming, and socially less dangerous than alcohol. An old and irrational prejudice against it keeps the "beat" and the "square" (square generally means any person who disapproves of jazz and/or marijuana; another opposite of "square" is "hip" from "hipster"—one who is knowledgeable around drugs and jazz) on different sides of the fence, and the beat is often on the inner side of a jail fence in this case. Now for "irresponsible." The irresponsible people in this world are the generals and politicians who test nuclear bombs; the antisocial, violent, and childish people are the ones who are running the world's governments right now. To refuse to participate in their idiocy—and this means keeping out of jobs that contribute to military preparations, staying out of the army, and saying what you think without fear of anyone—is a real responsibility and one for poets to face up to.

In a way one can see the beat generation as another aspect of the perpetual "third force" that has been moving through history with its own values of community, love, and freedom. It can be linked with the ancient Essene communities, primitive Christianity, Gnostic communities, and the free-spirit heresies of the Middle Ages; with Islamic Sufism, early Chinese Taoism, and both Zen and Shin Buddhism. The bold and moving erotic sculptures at Konorak in India, the paintings of Hieronymus Bosch, the poetry of William Blake, all belong to the same tradition. The motto in a Los Angeles beat coffeehouse is the equation "Art Is Love Is God." In America we get this through Walt Whitman and Henry David Thoreau and from our teachers of the generation above us, William Carlos Williams, Robinson Jeffers, Kenneth Rexroth, Henry Miller, and D. H. Lawrence.

What is the international significance of all this? The beat generation can be seen as an aspect of the worldwide trend for intellectuals to reconsider the nature of the human individual, existence, personal motives, the qualities of love and hatred, and the means of achieving wisdom. Existentialism, the modern pacifist-anarchist movement, the current interest of Occidentals in Zen Buddhism, are all a part of that trend. The beat generation is particularly interesting because it is not an intellectual movement, but a creative one: people who have cut their ties with respectable society in order to live an independent way of life writing poems and painting pictures, making mistakes and taking chances—but finding no reason for apathy or discouragement. They are going somewhere. It would do no harm if some of their attitudes came to liven up the poets of Japan.

THE NEW WIND

Everyone has heard now about the "poetry renaissance" and the "beat generation" in America (it spreads to Asia and Europe); and now, finally, a full-sized anthology of this poetry is in print: *The New American Poetry 1945–1960* edited by Donald Allen (Grove Press). I just received this book a few days ago, and it presents the poets of my generation so well that writing this article will be in a way a review of Don Allen's book.

What is "new" about the new American poetry? First,

what is new about the poets? The most striking thing is their detachment from the official literary world, be it publishing and commercial magazines or the literature departments of universities. They earn their livings in a wide variety of ways, but feel their real work to be poetry—requiring no justification. They are well educated, tough, independent, and in no way lumped into cliques. They have kept out (or been kept out) of the comfortable middle-class life in America. Many of them consider the universities to be instituted for professional liars and call them "fog factories"—and feel they are doing truth a service by staying in a position to teach and say what they like. They are gathered largely in San Francisco and New York, but a lot of them hide out in the country; some are even farmers. Places like Big Sur on the California coast (where Henry Miller has lived for over a decade) attract quite a few. They are different from their immediate predecessors in this detachment from the universities and in the fact that they have rejected the academic and neoformalist poetry of the late thirties and forties. *Poetry* magazine, which published Ezra Pound, T. S. Eliot, and a host of other now-famous modern poets in the twenties, is now considered conservative and prints almost none of the people in *The New American Poetry*.

They take poetry very seriously and have some striking views: poetry is not "beauty," not propaganda for the Bolsheviks or capitalists or whatever, not drinking sake and enjoying the moonlight, not "recollection in tranquillity" (Wordsworth), but a combination of the highest activity of trained intellect and the deepest insight of the intuitive, instinctive, or emotional mind, "all the faculties"—for Mike McClure it is "protein and fire"—it is in breath (Charles Olson) making the formal line, a sort of stylistic *ch'i*. It is sensitive awareness to things as they are (which if overstressed, takes the guts and brains out of poetry, as has too often happened in Japan), it is history, and most of all it is Magic, the power to transform by symbol and metaphor, to create a world with forms or to destroy a world with chaos. Charles Olson, in his note (at the back of the anthology) on "Projective Verse," says *form* is an extension of *content*. There is a philosophical paradigm in Buddhism that runs parallel to this and exposes the whole nature of the form/content dualism: form = emptiness, emptiness = form.

So: this poetry is in the "modern" tradition—harking back to Ezra Pound and particularly William Carlos Williams (who has been the largest single influence on the present generation of writers) but skipping over the poets of the late thirties and forties, who ran the influential poetry magazines, as being too formal and too dull. There is certainly nothing dull about the new American poetry—it is controversial, direct, topical, uninhibited, based on personal experience. The poets in this collection most closely identified with the "beat generation" are simply the most erratically imaginative, "crazy," and uninhibited. But not all the new poets would like to be described as members of the beat generation.

The nineteenth-century American poets Whitman and Herman Melville (whose fame as a poet, apart from his novels, is quite recent) have been highly appreciated. From Europe, the poetry in Spanish of Rafael Alberti and Federico García Lorca (especially!); the French poets Henri Michaux,

Antonin Artaud, Guillaume Apollinaire; the German Bertolt Brecht—and Pablo Neruda of Chile—have been influential. The classical Chinese poets through the translations of Ezra Pound, Arthur Waley, and Kenneth Rexroth; the translations of Japanese haiku by R. H. Blyth, a rereading of the Greek anthology and the medieval European goliard poets have all come together to produce a poetry of irony and compassion.

The fact is there is a creative flowering in America right now the like of which we haven't seen before. Poets are everywhere, and most of them are pretty good. One is tempted to say, like Elizabethan England or the mid T'ang dynasty. It is astounding.

Don Allen, in his preface to *The New American Poetry*, divides the poets into five classes, which I will list here in case it helps anyone understand.

1. The poets who were first associated with *Origin* magazine (founded by Cid Corman, who lived quietly in Kyoto for two years ending April 1960) and *Black Mountain Review*. The most outstanding of these are Charles Olson and Robert Duncan. Others are Denise Levertov, Paul Blackburn, Robert Creeley, Paul Carroll, Larry Eigner, Edward Dorn, Jonathan Williams, and Joel Oppenheimer. These people are writing a very careful, competent, and wise poetry.

2. The San Francisco group. Kenneth Rexroth, who is too old to be included in this anthology, is in a sense the father of the San Francisco poetry renaissance, along with Robert Duncan (who somehow ended up in class 1). The poets represented are Helen Adam, Brother Antoninus

(William Everson), James Broughton, Madeline Gleason, Lawrence Ferlinghetti, Robin Blaser, Jack Spicer, Lew Welch, Richard Duerden, Philip Lamantia, Bruce Boyd, Kirby Doyle, and Ebbe Borregaard. Ferlinghetti is the owner of City Lights Bookshop in San Francisco, which first published Ginsberg's *Howl*.

3. The beat generation. Don Allen strictly defines it to include only Jack Kerouac, Allen Ginsberg, Gregory Corso, and Peter Orlovsky. Kerouac is best known as a novelist, but he has a delightful book of poems in print (Grove Press) called *Mexico City Blues*—poems to a large extent concerned with Kerouac's personal interpretation of Buddhism.

4. New York poets. Those included are Barbara Guest, James Schuyler, Edward Field, Kenneth Koch, Frank O'Hara, and John Ashbery. Frank O'Hara is probably the best known.

5. The last group is a sort of miscellaneous collection of independent characters who cannot be fitted into any other class and who all have individual styles. They are Philip Whalen, Gilbert Sorrentino, Stuart Perkoff, Edward Marshall, Michael McClure, Ray Bremser, LeRoi Jones (editor of *Yugen* magazine), John Weiners, Ron Loewinsohn, David Meltzer, and myself. Actually most of these poets could be called San Francisco people, but for some reason Don Allen prefers to keep them separate.

The New American Poetry ends up with "Statements on Poetics" by various poets, biographical notes on everybody, and a short bibliography.

This book will become the handbook of young writers in

the sixties. It is a good selection. I can only think of three poets I would like to see in it who aren't (Cid Corman, by his own choice; Theodore Enslin; and Tom Parkinson). Let us hope this fresh wind of poetry doesn't stay just in America, but blows its way about the globe.

[These two pieces were published in Japanese in Chuo-koron, *a highly regarded Japanese intellectual journal, in 1960. I was living then in Kyoto. I wrote them at the urging of Hisao Kanaseki, a lifelong mentor on transpacific literary matters, who wanted me to introduce the beat generation and the new American poetry to the urban intelligentsia. These folks were existentialist Marxists, with a French symbolist aesthetic—an imported mindscape, incompletely assimilated into the just-beginning Japanese industrial renaissance and rising affluence. These pieces reflect the downright joy I felt over the new American poetry. They were first published in English in* American Poetry 2, *no. 1 (fall 1984).]*

A Virus Runs Through It

—Nothing is true—Everything is permitted—
"*No hassan i sabbah—we want flesh—we want
junk—we want power—*"
"That did it—Dial *police*"—

WILLIAM S. BURROUGHS
The Ticket That Exploded (1962)

*T*he "ticket" is Western culture plus "civilization"—the world's dominant sort of society for the last five thousand years. Humans have been hoping for some time that it would take them somewhere. Now it is exploding in our hands.

> " '*Mr Bradly Mr Martin*' also known as '*the Ugly Spirit,*' *thought to be the leader of the mob—the nova mob—In all my experience as a police officer i have never seen such total fear and degradation on any planet—We intend to arrest these criminals and turn them over to the Biological Department for the indicated alterations—*"

Under the guise of literature or science-fiction fantasy, Mr. William Burroughs is writing down about as clear as it can be writ an outline of the present cultural, spiritual, and political struggle that is taking place not in any obvious area but in the consciousness of beings on this planet. Cosmic politics: politics of the nervous system. The book is too complex to plunder for demonstrations of "ideas." Burroughs's writing techniques here derive from earlier cut-up experiments and now employ sophisticated mixing and scrambling of audiotapes. The language is a hyped-up criminal-vaudeville-carny tough-cop hip talk resonating with "scientific authority" and a whiff of con. The book makes the reader a bit nervous.

The story line, like a Hindu Purana, recounts the destruction and renovation of a universe. The human realm becomes a temporary battleground for forces from "outer space." The conflict is not between simple powers of "light" and "darkness," as Burroughs takes pains to make clear, but between agents of unbalance, in the form of overreaching craving and aggression, and an integrating, balancing force. Thus Vishnu took human form as Krishna to subdue the tyrant Kansa—not as any final victory over "evil" but as a necessary restoration of harmonies so that cowherds could work out their karma without undue obstruction.

The nova mob enjoys aggravating conflicts on planets by encouraging the growth of mutually incompatible life forms. The payoff is the explosion of the planet: it becomes a nova. Nova criminals "are not three-dimensional organisms" but like a virus, they require a human host. They are able to enter the host through a style, vice, or habit that is their trademark. "We were able to trace Hamburger Mary through her fondness for peanut butter" as this particular criminal passed from host to host. "Some move on junk lines through addicts of the earth, others move on lines of certain sexual practices and so forth—It is only when we can block the controller out of all coordinate points available to him and flush him out from host cover that we can make a definitive arrest—Otherwise the criminal escapes to other coordinates—"

The nova criminal most immediately responsible for the situation on earth is "Mr Bradly Mr Martin" who works through junk (he is actually "a heavy metal addict from Uranus") and ugliness. From his office on another planet, Inspector J. Lee of the nova police works to capture or isolate these criminals—send them back for biological alterations—send them back to "Rewrite." Reality is open to re-editing.

At the heart of the story is Mr. Burroughs's view of the nervous system. In many ways we are not separate entities; we are all sharing, living in, the same nervous system. Different societies have various methods of controlling and manipulating consciousness, starting with sex and language. Sexual behavior and habit, body chemistry, thought and image association lines become more controllable as civilized societies develop increasingly sophisticated communication and power systems. (Imperialism can perhaps be seen as a culture's ego paranoia and overreaching anxiety growth. It may well be that part of humanity's present angst is a deep "species guilt"—a profound inward awareness that human-kind itself has become a tyrannical and destructive breed in the biological scheme and is ripe for correction.)

At times Mr. Burroughs almost opts for Total Liberation as he puts down even the basic structure of higher life forms on this planet: "The human organism is literally consisting of two halves from the beginning word and all human sex is this unsanitary arrangement whereby two entities attempt to occupy the same three-dimensional coordinate points giving rise to the sordid latrine brawls which have characterized a planet based on 'the Word,' that is, on separate flesh engaged in endless sexual conflict—"

Contemporary society is witnessing (via electronic media) the proliferation of nervous-system extensions to such a degree that political struggles today are in essence campaigns to gain the "mind" of some other group. The mass media (for Burroughs the tape recorder is a key tool) are indeed an exteriorized nervous system: a planet-circling electric social consciousness where a delighted laugh or a wrenching groan can ripple across populations within minutes. In *The Ticket That Exploded*, sex (ejaculations and orgasms) move like points of light flashing at coordinate points, sweeping through millions of specialized "flesh addicts." *Addiction* is a key term. Consumer craving whipped up by advertising: alcohol, tobacco, phony sex (a section of the book is called "substitute flesh": "Better than the real thing!—There is no real thing—Maya—Maya—It's all show business"). Chemicals, warm-water sensory-deprivation experiments, Wilhelm Reich's orgone theory are all briefly scanned into the story. The pivotal twist lies in the media and its manipulation by Mr Bradley Mr Martin, who is in the pay of—whom? Corporation board chairs or professional politicians who still believe humanity was meant to be the Lord of Creation, hence individuals must be kept busy working? Aesthete fascist megalomaniacs? The whole process leads to ultimate sadomasochistic TV extravaganzas—"orgasm *is* death"—and some discussion of the Hanged Man (who was important to *Naked Lunch*, too).

It's also a story of "theories" and swims in a rhetoric of "science," as Burroughs tells of experiments that trick the nervous system with audiotape and film splicing, editing, cutting up. A brief description is suggestive: Burroughs's narrative voice tells how in splicing two voices onto one tape—alternating the voices at very short intervals, and then playing them back (if my understanding is correct)—one gets a remarkable breakdown of separate entities, and a third, new entity emerges whose existence is in the medium of sound alone. This new personage, Burroughs says, is real: at various speeds and with sudden start-stop playback maneuvers, you can hear words being spoken that aren't on the original tapes. The effect is unsettling and "erotic." The practical effect on personality, we are told, is enlightening, since it destroys one's usual habits of thought-and-word association. For modern humans, word associating has become "obsessional"; it also leaves one open to control from outside. So, Burroughs suggests, an extensive audiotape spiritual-psychological discipline will liberate you from your old controlled nervous-system habits better than "twenty years of sitting in the lotus posture."

Ticket lays bare a number of cosmic psycho-tricks and

suggests how they work on us, with some remarkable theories and suggestions that are the madness verging on truth. It is a contemporary irony that the hero (if any) is one J. Lee, the dryly professional (interplanetary) police inspector. We know that the police on this earth are usually emanations of hysteria and anger and not an expression of a mature social wisdom, but we still have a wistful hope that incorruptible warriors of the Dharma might be out there somewhere, able to understand and help us poor human beings who have unwittingly become host to so much madness. Bill Burroughs's insistent quack-doctor-shaman voice and his dubious bitter medicine are very likely good for you.

[This review was written in 1962 and has not been previously published. It's here now because Burroughs's book, with its evocation of the politics of addiction, mass madness, and virus panic, is all too prophetic.]

Smokey the Bear Sutra

*O*nce in the Jurassic, about 150 million years ago, the Great Sun Buddha in this corner of the Infinite Void gave a great Discourse to all the assembled elements and energies: to the standing beings, the walking beings, the flying beings, and the sitting beings—even the grasses, to the number of thirteen billion, each one born from a seed—assembled there: a Discourse concerning Enlightenment on the planet Earth.

"In some future time, there will be a continent called America. It will have great centers of power such as Pyramid Lake, Walden Pond, Mount Rainier, Big Sur, the Everglades, and so forth, and powerful nerves and channels such as the Columbia River, Mississippi River, and Grand Canyon. The human race in that era will get into troubles all over its head and practically wreck everything in spite of its own strong intelligent Buddha-nature.

"The twisting strata of the great mountains and the pulsings of great volcanoes are my love burning deep in the earth. My obstinate compassion is schist and basalt and granite, to be mountains, to bring down the rain. In that future

American Era I shall enter a new form, to cure the world of loveless knowledge that seeks with blind hunger, and mindless rage eating food that will not fill it."

And he showed himself in his true form of

SMOKEY THE BEAR.

A handsome smokey-colored brown bear standing on his hind legs, showing that he is aroused and watchful.

Bearing in his right paw the Shovel that digs to the truth beneath appearances, cuts the roots of useless attachments, and flings damp sand on the fires of greed and war;

His left paw in the Mudra of Comradely Display—indicating that all creatures have the full right to live to their limits and that deer, rabbits, chipmunks, snakes, dandelions, and lizards all grow in the realm of the Dharma;

Wearing the blue work overalls symbolic of slaves and laborers, the countless people oppressed by a civilization that claims to save but only destroys;

Wearing the broad-brimmed hat of the West, symbolic of the forces that guard the Wilderness, which is the Natural State of the Dharma and the True Path of beings on earth— all true paths lead through mountains—

With a halo of smoke and flame behind, the forest fires of the kali yuga, fires caused by the stupidity of those who think things can be gained and lost whereas in truth all is contained vast and free in the Blue Sky and Green Earth of One Mind;

Round-bellied to show his kind nature and that the great Earth has food enough for everyone who loves her and trusts her;

Trampling underfoot wasteful freeways and needless suburbs; smashing the worms of capitalism and totalitarianism;

Indicating the Task: his followers, becoming free of cars, houses, canned food, universities, and shoes, master the Three Mysteries of their own Body, Speech, and Mind, and fearlessly chop down the rotten trees and prune out the sick limbs of this country America and then burn the leftover trash.

Wrathful but Calm, Austere but Comic, Smokey the Bear will illuminate those who would help him; but for those who would hinder or slander him,

HE WILL PUT THEM OUT.

Thus his great Mantra:

Namah samanta vajranam chanda maharoshana
Sphataya hum traka ham mam

"I DEDICATE MYSELF TO THE UNIVERSAL DIAMOND—
BE THIS RAGING FURY DESTROYED"

And he will protect those who love woods and rivers, Gods and animals, hoboes and madmen, prisoners and sick people, musicians, playful women, and hopeful children;

And if anyone is threatened by advertising, air pollution, or the police, they should chant SMOKEY THE BEAR'S WAR SPELL:

DROWN THEIR BUTTS
CRUSH THEIR BUTTS

DROWN THEIR BUTTS

CRUSH THEIR BUTTS

And SMOKEY THE BEAR will surely appear to put the enemy out with his vajra shovel.

Now those who recite this Sutra and then try to put it
in practice will accumulate merit as countless as the
sands of Arizona and Nevada,
Will help save the planet Earth from total oil slick,
Will enter the age of harmony of humans and nature,
Will win the tender love and caresses of men, women,
and beasts,
Will always have ripe blackberries to eat and a sunny
spot under a pine tree to sit at,

AND IN THE END WILL WIN HIGHEST PERFECT
ENLIGHTENMENT.

Thus have we heard.

(may be reproduced free forever)

REGARDING "SMOKEY THE BEAR SUTRA"

It's hard not to have a certain amount of devotional feeling for the Large Brown Ones, even if you don't know much about them. I met the Old Man in the Fur Coat a few times in the North Cascades—once in the central Sierra—and was suitably impressed. There are many stories told about humans marrying the Great Ones. I brought much of that lore together in my poem "this poem is for B___r," which is part of *Myths and Texts*. The Circumpolar B___r cult, we are told, is the surviving religious complex (stretching from Suomi to Utah via Siberia) of what may be the oldest religion on earth. Evidence in certain Austrian caves indicates that our Neanderthal ancestors were practicing a devotional ritual to the Big Fellow about seventy thousand years ago. In the light of meditation once it came to me that the Old One was no other than that Auspicious Being described in Buddhist texts as having taught in the unimaginably distant past, the one called "The Ancient Buddha."

So I came to realize that the U.S. Forest Service's "Smokey the B___r" publicity campaign was the inevitable resurfacing of our ancient benefactor as guide and teacher in the twentieth century, the agency not even knowing that it was serving as a vehicle for this magical reemergence.

During my years in Japan I had kept an eye out for traces of ancient B___r worship in folk religion and within Buddhism, and it came to me that Fudo Myoō, the patron of the Yamabushi (a Shinto-Buddhist society of mountain yogins), whose name means the "Immovable Wisdom King," was possibly one of those traces. I cannot provide an academic proof for this assertion; it's an intuition based on Fudo's usual habitat: deep mountains. Fudo statues and paintings portray a wickedly squinting fellow with one fang down and one fang up, a braid hanging down one side of the head, a funny gleam in his eye, wreathed in rags, holding a vajra sword and a lariat, standing on rough rock and surrounded by flames. The statues are found by waterfalls and deep in the wildest mountains of Japan. He also lurks in caves. Like the Ainu's Kamui Kimun, Lord of the Inner Mountains— clearly a B__r deity—Fudo has surpassing power, the capacity to quell all lesser violence. In the iconography he is seen as an aspect of Avalokiteśvara, the Bodhisattva of Compassion, or the consort of the beautiful Bodhisattva Tārā, She Who Saves.

It might take this sort of Buddha to quell the fires of greed and war and to help us head off the biological holocaust that the twenty-first century may well prove to be. I had such thoughts in mind when I returned to Turtle Island (North America) in December of 1968 from a long stay in Japan. A copy of the *San Francisco Chronicle* announced the Sierra Club Wilderness conference of February 1969; it was to be the following day. I saw my chance, sat down, and the sutra seemed to write itself. It follows the structure of a Mahayana Buddhist sutra fairly faithfully. The power mantra of the Great Brown One is indeed the mantra of Fudo the Immovable.

I got it printed overnight. The next morning I stood in the lobby of the conference hotel in my old campaign hat and handed out the broadsides, saying, "Smokey the B__r literature, sir." Bureau of Land Management and Forest Service officials politely took them. Forest beatniks and conservation fanatics read them with mad glints and giggles. The Underground News Service took it up, and it went to the *Berkeley Barb* and then all over the country. *The New Yorker* queried me about it, and when I told them it was both free and anonymous, they said they couldn't publish it. It soon had a life of its own, as intended.

[This commentary was written to accompany the sutra's inclusion in the anthology Working the Woods, Working the Sea, *edited by Finn Wilcox and Jeremiah Gorsline (Port Townsend, Wash.: Empty Bowl Press, 1986).]*

Four Changes, with a Postscript

I. Population

The Condition

POSITION: Human beings are but a part of the fabric of life—dependent on the whole fabric for their very existence. As the most highly developed tool-using animal, we must recognize that the unknown evolutionary destinies of other life forms are to be respected, and we must act as gentle steward of the earth's community of being.

SITUATION: There are now too many human beings, and the problem is growing rapidly worse. It is potentially disastrous not only for the human race but for most other life forms.

GOAL: The goal would be half of the present world population or less.

Action

SOCIAL/POLITICAL: First, a massive effort to convince the governments and leaders of the world that the problem is severe. And that all talk about raising food production—well intentioned as it is—simply puts off the only real solution: reduce population. Demand immediate participation by all countries in programs to legalize abortion, encourage vasectomy and sterilization (provided by free clinics); try to correct traditional cultural attitudes that tend to force women into childbearing; remove income-tax deductions for more than two children above a specified income level, and scale it so that lower-income families are forced to be careful, too, or pay families to limit their number. Take a vigorous stand against the policy of the right wing in the Catholic hierarchy and any other institutions that exercise an irresponsible social force in regard to this question; oppose and correct simpleminded boosterism that equates population growth with continuing prosperity. Work ceaselessly to have all political questions be seen in the light of this prime problem.

In many cases, the governments are the wrong agents to address. Their most likely use of a problem, or crisis, is as another excuse for extending their own powers. Abortion should be legal and voluntary. Great care should be taken that no one is ever tricked or forced into sterilization. The whole population issue is fraught with contradictions, but the fact stands that by standards of planetary biological welfare there are already too many human beings. The long-range answer is a steady low birthrate. Area by area of the globe, the measure of "optimum population" should be based on what is best for the total ecological health of the region, including its wildlife populations.

THE COMMUNITY: Explore other social structures and marriage forms, such as group marriage and polyandrous marriage, which provide family life but many less children. Share the pleasures of raising children widely, so that all need not directly reproduce in order to enter into this basic human experience. We must hope that no woman would give birth to more than one or two children during this period of crisis. Adopt children. Let reverence for life and reverence for the feminine mean also a reverence for other species and for future human lives, most of which are threatened.

OUR OWN HEADS: "I am a child of all life, and all living beings are my brothers and sisters, my children and grandchildren. And there is a child within me waiting to be born, the baby of a new and wiser self." Love, lovemaking, seen as the vehicle of mutual realization for a couple, where the creation of new selves and a new world of being is as important as reproducing our kind.

II. POLLUTION

The Condition
POSITION: Pollution is of two types. One sort results from an excess of some fairly ordinary substance—smoke, or solid waste—that cannot be absorbed or transmitted rapidly enough to offset its introduction into the environment, thus

causing changes the great cycle is not prepared for. (All organisms have wastes and by-products, and these are indeed part of the total biosphere: energy is passed along the line and refracted in various ways. This is cycling, not pollution.) The other sort consists of powerful modern chemicals and poisons, products of recent technology that the biosphere is totally unprepared for. Such are DDT and similar chlorinated hydrocarbons; nuclear testing fallout and nuclear waste; poison gas, germ and virus storage and leakage by the military; and chemicals that are put into food, whose long-range effects on human beings have not been properly tested.

SITUATION: The human race in the last century has allowed its production and scattering of wastes, by-products, and various chemicals to become excessive. Pollution is directly harming life on the planet—which is to say, ruining the environment for humanity itself. We are fouling our air and water and living in noise and filth that no "animal" would tolerate, while advertising and politicians try to tell us we've never had it so good. The dependence of modern governments on this kind of untruth leads to shameful mind pollution, through the mass media and much school education.

GOAL: Clean air, clean clear-running rivers; the presence of pelican and osprey and gray whale in our lives; salmon and trout in our streams; unmuddied language and good dreams.

Action

SOCIAL/POLITICAL: Effective international legislation banning DDT and other poisons—with no fooling around. The collusion of certain scientists with the pesticide industry and agribusiness in trying to block this legislation must be brought out in the open. Strong penalties for water and air pollution by industries: Pollution is somebody's profit. Phase out the internal combustion engine and fossil fuel use in general; do more research into nonpolluting energy sources, such as solar energy, the tides. No more kidding the public about nuclear waste disposal: it's impossible to do it safely, so nuclear-generated electricity cannot be seriously planned for as it now stands. Stop all germ and chemical warfare research and experimentation; work toward a safe disposal of the present stupid and staggering stockpiles of H-bombs, cobalt gunk, germ and poison tanks and cans. Provide incentives against the wasteful use of paper and so on, which adds to the solid wastes of cities—develop methods of recycling solid urban wastes. Recycling should be the basic principle behind all waste-disposal thinking. Thus, all bottles should be reusable; old cans should make more cans; old newspapers should go back into newsprint again. Establish stronger controls and conduct more research on chemicals in foods. A shift toward a more varied and sensitive type of agriculture (more small-scale and subsistence farming) would eliminate much of the call for the blanket use of pesticides.

THE COMMUNITY: DDT and such: don't use them. Air pollution: use fewer cars. Cars pollute the air, and one or two people riding lonely in a huge car is an insult to intelligence and the earth. Share rides, legalize hitchhiking, and build hitchhiker waiting stations along the highways. Also— a step toward the new world—walk more; look for the best routes through beautiful countryside for long-distance walking trips: San Francisco to Los Angeles down the Coast Range, for example. Learn how to use your own manure as fertilizer if you're in the country, as people in the Far East have done for centuries. There's a way, and it's safe. Solid waste: boycott bulky wasteful Sunday papers, which use up trees. It's all just advertising anyway, which is artificially inducing more energy consumption. Refuse bags at the store and bring your own. Organize park and street cleanup festivals. Don't work in any way for or with an industry that pollutes, and don't be drafted into the military. Don't waste. (A monk and an old master were once walking in the mountains. They noticed a little hut upstream. The monk said, "A wise hermit must live there." The master said, "That's no wise hermit—you see that lettuce leaf floating down the stream? He's a Waster." Just then an old man came running down the hill with his beard flying and caught the floating lettuce leaf.) Carry your own jug to the winery and have it filled from the barrel.

OUR OWN HEADS: Part of the trouble with talking about something like DDT is that the use of it is not just a practical device, it's almost an establishment religion. There is something in Western culture that wants to wipe out creepy-crawlies totally and feels repugnance for toadstools and snakes. This is fear of one's own deepest inner-self wilderness areas, and the answer is *relax*. Relax around bugs, snakes, and your own hairy dreams. Again, we all should share our crops with a certain percentage of bug life as a way of "paying our dues." Thoreau says, "How then can the harvest fail? Shall I not rejoice also at the abundance of the weeds whose seeds are the granary of the birds? It matters little comparatively whether the fields fill the farmer's barns. The true husbandman will cease from anxiety, as the squirrels manifest no concern whether the woods will bear chestnuts this year or not, and finish his labor with every day, relinquish all claim to the produce of his fields, and sacrificing in his mind not only his first but his last fruits also." In the realm of thought, inner experience, consciousness, as in the outward realm of interconnection, there is a difference between balanced cycle and the excess that cannot be handled. When the balance is right, the mind recycles from highest illuminations to the muddy blinding anger or grabbiness that sometimes seizes us all—the alchemical "transmutation."

III. Consumption

The Condition

POSITION: Everything that lives eats food and is food in turn. This complicated animal, the human being, rests on a vast and delicate pyramid of energy transformations. To grossly use more than you need, to destroy, is biologically unsound. Much of the production and consumption of modern societies is not necessary or conducive to spiritual and cultural growth, let alone survival, and is behind much greed and envy, age-old causes of social and international discord.

SITUATION: Humanity's careless use of "resources" and its total dependence on certain substances such as fossil fuels (which are being exhausted, slowly but certainly) are having harmful effects on all the other members of the life network. The complexity of modern technology renders whole populations vulnerable to the deadly consequences of the loss of any one key resource. Instead of independence we have over-dependence on life-giving substances such as water, which we squander. Many species of animals and birds have become extinct in the service of fashion fads, or fertilizer, or industrial oil. The soil is being used up; in fact, humanity has become a locustlike blight on the planet that will leave a bare cupboard for its own children—all the while living in a kind of addict's dream of affluence, comfort, eternal progress, using the great achievements of science to produce software and swill.

GOAL: Balance, harmony, humility, growth that is a mutual growth with redwood and quail; to be a good member of the great community of living creatures. True affluence is not needing anything.

Action

SOCIAL/POLITICAL: It must be demonstrated ceaselessly that a continually "growing economy" is no longer healthy, but a cancer. And that the criminal waste that is allowed in the name of competition—especially that ultimate in wasteful needless competition, hot wars and cold wars with "communism" (or "capitalism")—must be halted totally with ferocious energy and decision. Economics must be seen as a small subbranch of ecology, and production/distribution/consumption handled by companies or unions or cooperatives with the same elegance and spareness one sees in nature. Soil banks; open spaces; logging to be truly based on sustained yield (the U.S. Forest Service is—sadly—now the lackey of business). Protection for all scarce predators and varmints: "Support your right to arm bears." Damn the International Whaling Commission, which is selling out the last of our precious, wise whales; ban absolutely all further development of roads and concessions in national parks and wilderness areas; build auto campgrounds in the least desirable areas. Initiate consumer boycotts of dishonest and unnecessary products. Establish co-ops. Politically, blast both "communist" and "capitalist" myths of progress and all crude notions of conquering or controlling nature.

THE COMMUNITY: Sharing and creating. The inherent aptness of communal life, where large tools are owned jointly and used efficiently. The power of renunciation: if enough Americans refused to buy a new car for one given year, it would permanently alter the American economy. Recycling clothes and equipment. Support handicrafts, gardening, home skills, midwifery, herbs—all the things that can make us independent, beautiful, and whole. Learn to break the habit of acquiring unnecessary possessions—a monkey on everybody's back—but avoid a self-abnegating antijoyous self-righteousness. Simplicity is light, carefree, neat, and loving—not a self-punishing ascetic trip. (The great Chinese poet Tu Fu said, "The ideas of a poet should be noble and simple.") Don't shoot a deer if you don't know how to use all the meat and preserve that which you can't eat, to tan the hide and use the leather—to use it all, with gratitude, right down to the sinew and hooves. Simplicity and mindfulness in diet are the starting point for many people.

OUR OWN HEADS: It is hard even to begin to gauge how much a complication of possessions, the notions of "my and mine," stand between us and a true, clear, liberated way of seeing the world. To live lightly on the earth, to be aware and alive, to be free of egotism, to be in contact with plants and animals, starts with simple concrete acts. The inner principle is the insight that we are interdependent energy fields of great potential wisdom and compassion, expressed in each person as a superb mind, a handsome and complex body, and the almost magical capacity of language. To these potentials and capacities, "owning things" can add nothing of authenticity. "Clad in the sky, with the earth for a pillow."

IV. TRANSFORMATION

The Condition

POSITION: Everyone is the result of four forces: the conditions of this known universe (matter/energy forms and ceaseless change), the biology of his or her species, individual genetic heritage, and the culture one is born into. Within this web of forces there are certain spaces and loops that allow to some persons the experience of inner freedom and illumination. The gradual exploration of some of these spaces constitutes "evolution" and, for human cultures, what "history" could increasingly be. We have it within our deepest powers not only to change our "selves" but to change our culture. If humans are to remain on earth, they must transform the five-millennia-long urbanizing civilization tradition into a new ecologically sensitive harmony-oriented wild-minded scientific-spiritual culture. "Wildness is the state of complete awareness. That's why we need it."

SITUATION: Civilization, which has made us so successful a species, has overshot itself and now threatens us with its inertia. There is also some evidence that civilized life isn't good for the human gene pool. To achieve the Changes, we must change the very foundations of our society and our minds.

GOAL: Nothing short of total transformation will do much good. What we envision is a planet on which the human population lives harmoniously and dynamically by employing various sophisticated and unobtrusive technologies in a world environment that is "left natural." Specific points in this vision:

- A healthy and spare population of all races, much less in number than today.
- Cultural and individual pluralism, unified by a type of world tribal council. Division by natural and cultural boundaries rather than arbitrary political boundaries.
- A technology of communication, education, and quiet transportation, land use being sensitive to the properties of each region. Allowing, thus, the bison to return to much of the High Plains. Careful but intensive agriculture in the great alluvial valleys; deserts left wild for those who would live there by skill. Computer technicians who run the plant part of the year and walk along with the elk in their migrations during the rest.
- A basic cultural outlook and social organization that inhibits power and property seeking while encouraging exploration and challenge in things like music, meditation, mathematics, mountaineering, magic, and all other ways

of authentic being-in-the-world. Women totally free and equal. A new kind of family—responsible, but more festive and relaxed—is implicit.

Action

SOCIAL/POLITICAL: It seems evident that there are throughout the world certain social and religious forces that have worked through history toward an ecologically and culturally enlightened state of affairs. Let these be encouraged: Gnostics, hip Marxists, Teilhard de Chardin Catholics, Druids, Taoists, Biologists, Witches, Yogins, Bhikkus, Quakers, Sufis, Tibetans, Zens, Shamans, Bushmen, American Indians, Polynesians, Anarchists, Alchemists—the list is long. Primitive cultures, communal and ashram movements, cooperative ventures. Since it doesn't seem practical or even desirable to think that direct bloody force will achieve much, it would be best to consider this change a continuing "revolution of consciousness," which will be won not by guns but by seizing the key images, myths, archetypes, eschatologies, and ecstasies so that life won't seem worth living unless one's on the side of the transforming energy. We must take over "science and technology" and release its real possibilities and powers in the service of this planet—which, after all, produced us and it.

(More concretely: no transformation without our feet on the ground. Stewardship means, for most of us, find your place on the planet, dig in, and take responsibility from there—the tiresome but tangible work of school boards, county supervisors, local foresters, local politics, even while holding in mind the largest scale of potential change. Get a sense of workable territory, learn about it, and start acting point by point. On all levels, from national to local, the need to move toward steady state economy—equilibrium, dynamic balance, inner growth stressed—must be taught. Maturity/diversity/climax/creativity.)

THE COMMUNITY: New schools, new classes, walking in the woods and cleaning up the streets. Find psychological techniques for creating an awareness of "self" that includes the social and natural environment. Consideration of what specific language forms—symbolic systems—and social institutions constitute obstacles to ecological awareness. Without falling into a facile interpretation of McLuhan, we can hope to use the media. Let no one be ignorant of the facts of biology and related disciplines; bring up our children as part of the wildlife. Some communities can establish themselves in backwater rural areas and flourish, others can maintain themselves in urban centers, and the two types can work together—a two-way flow of experience, people, money, and homegrown vegetables. Ultimately cities may exist only as joyous tribal gatherings and fairs, to dissolve after a few weeks. Investigating new lifestyles is our work, as is the exploration of ways to explore our inner realms—with the known dangers of crashing that go with such. Master the archaic and the primitive as models of basic nature-related cultures—as well as the most imaginative extensions of science—and build a community where these two vectors cross.

OUR OWN HEADS: Are where it starts. Knowing that we are the first human beings in history to have so much of our past culture and previous experience available to our study, and being free enough of the weight of traditional cultures to seek out a larger identity; the first members of a civilized society since the Neolithic to wish to look clearly into the eyes of the wild and see our selfhood, our family, there. We have these advantages to set off the obvious disadvantages of being as screwed up as we are—which gives us a fair chance to penetrate some of the riddles of ourselves and the universe, and to go beyond the idea of "human survival" or "survival of the biosphere" and to draw our strength from the realization that at the heart of things is some kind of serene and ecstatic process that is beyond qualities and beyond birth and death. "No need to survive!" "In the fires that destroy the universe at the end of the kalpa, what survives?" "—The iron tree blooms in the void!"

Knowing that nothing need be done is the place from which we begin to move.

Postscript (1995)

Four Changes was written in 1969. Michael McClure, Richard Brautigan, Steve Beckwitt, Keith Lampe, Cliff Humphreys, Alan Watts, Allen Hoffman, Stewart Brand, and Diane di Prima were among those who read it during its formative period and offered suggestions and criticisms. It was widely distributed in several free editions. I added a few more lines and comments in 1974, when it was published together with the poems in *Turtle Island* (New York: New Directions, 1974). Now it's 1995, and a quarter century has elapsed. The apprehension we felt in 1969 has not abated. It would be a fine thing to be able to say, "We were wrong. The natural world is no longer as threatened as we said then." One can take no pleasure, in this case, in having been on the right track. Many of the larger mammals face extinction, and all manner of species are endangered. Natural habitat ("raw land") is fragmented and then destroyed ("developed"). The world's forests are being relentlessly logged by multinational corporations. Air, water, and soil are all in worse shape. Population continues to climb, and even if it were a world of perfect economic and social justice, I would argue that ecological justice calls for fewer people. The few remaining traditional people with place-based sustainable economies are driven into urban slums and cultural suicide. The quality of life for everyone everywhere has gone down, what with resurgent nationalism, racism, violence both random and organized, and increasing social and economic inequality. There are whole nations for whom daily life is an ongoing disaster. Naive and utopian as some of it sounds now, I still stand by the basics of "Four Changes." As I wrote in 1969,

> My Teacher once said to me,
> —become one with the knot itself,
> 'til it dissolves away.
> —sweep the garden.
> —any size.

The Yogin and
the Philosopher

We live in a universe "one turn" in which, it is widely felt,
all are one and at the same time all are many. The extra
rooster and I were subject and object until one evening we
became one. As the discriminating, self-centered awareness
of civilized humans has increasingly improved their material survival potential, it has correspondingly moved them
further and further from a spontaneous feeling of being
part of the natural world. It often takes, ironically, an analytical and rational presentation from the biological sciences
of human interdependence with other life forms to move
modern people toward questioning their own role as major
planetary exploiter. This brings us to the use of terms like
"the rights of nonhuman nature" or questions such as "do
trees have standing?" From the standpoint of "all are one,"
the questions need never arise. The Chinese Buddhist
philosopher-monk Chan-jan proposed that even inanimate

things possess the Buddha-nature, saying, "The man who is of all-round perfection knows from beginning to end that no objects exist apart from Mind. Who then is 'animate' and who 'inanimate'? Within the Assembly of the Lotus, all are present without division."

From the standpoint of the seventies and eighties, it serves us well to examine the way we relate to these objects we take to be outside ourselves—nonhuman, nonintelligent, or whatever. If we are to treat the world (and ourselves) better, we must first ask, how can we know what the nonhuman realm is truly like? And second, if one gets a glimmer of an answer from there, how can it be translated, communicated, to the realm of humanity with its courts, congresses, and zoning laws? How do we listen? How do we speak?

The Cahuilla Indians who lived in the Palm Springs desert and the mountains above gathered plants from valley floor to mountain peak with precise knowledge. They said almost anybody can, if they pay enough attention and are patient, hear a little voice from plants. The Papago of southern Arizona said that a person who was humble and brave and persistent would some night hear a song in his dream, brought by the birds that fly in from the Gulf of California, or a hawk, a cloud, the wind, or the red rain spider, and the song would be his—would add to his knowledge and power.

How would one learn this sweet interspecies attention and patience? What practice would tune us in not just to dreams but to their *songs*? The philosopher speaks the language of reason, which is the language of public discourse, with the intention of being intelligible to anyone, without putting special demands on them apart from basic intelligence and education. Then there is religious discourse, involving acceptance of certain beliefs. There is also a third key style: the yogin. The yogin is an experimenter whose work brings forth a different sort of discourse, one of deep hearing and doing. The yogin experiments on herself. Yoga, from the root *yuj* (related to the English "yoke"), means to be at work, engaged. In India the distinction between philosopher and yogin is clearly and usefully made—even though sometimes the same individual might be both. The yogin has specific exercises, disciplines, by which she hopes to penetrate deeper in understanding than the purely rational function will allow. These practices, such as breathing, meditation, chanting, and so forth, are open to anyone to follow if one wishes, and the yogic traditions have long asserted that various people who have followed through a given course of practice usually came up with similar results. The yogins hold, then, that certain concepts of an apparently philosophical nature cannot actually be grasped except by proceeding through a set of disciplines. Thus the literature of the yogic tradition diverges from philosophical writing in that it makes special requirements of its readers. Note the difference between Plato and the school of Pythagoras. The latter was much closer to the style of India—ashrams, with special rules and dietary prohibitions. The alchemical, occult, Neoplatonic, and various sorts of Gnostic traditions of what might be called occidental counterphilosophy are strongly yogic in this way. Gnosticism took as its patroness Sophia, Wisdom, a goddess known in India under various names and in Buddhism under the name of Tārā, "She Who Saves" or leads across to the opposite shore. Witchcraft, a folk tradition going back to the Paleolithic, has its own associations of magic, feminine powers, and plant knowledge. As Robert Graves points out in *The White Goddess*, the convergence of many ancient religious and shamanistic lines produces the western lore of the muse. Some sorts of poetry are the mode of expression of certain yogic-type schools of practice. In fact, singing comes very close to being a sort of meditation in its own right; some recent research holds that song is a "right-hemisphere-of-the-brain" function, drawing on the intuitive, creative, nonverbal side of human consciousness. Since speech is a left-hemisphere function, poetry (word and song together) is surely a marriage of the two halves.

The philosopher and the poet-yogin both have standing not too far behind them the shaman, with his or her pelt and antlers or various other guises, and with songs going back to the Pleistocene and before. The shaman speaks for wild animals, the spirits of plants, the spirits of mountains, of watersheds. He or she sings for them. They sing through him. This capacity has been achieved via sensibilities and disciplines. In the shaman's world, wilderness and the unconscious become analogous: she who knows and is at ease in one will be at home in the other.

The elaborate, yearly, cyclical production of grand ritual dramas in the societies of Pueblo Indians of North America (for one example) can be seen as a process by which the whole society consults the nonhuman (in-human, inner-human?) powers and allows some individuals to step totally out of their human roles to put on the mask, costume, and *mind* of Bison, Bear, Squash, Corn, or Pleiades; to reenter the human circle in that form and by song, mime, and dance, convey a greeting from the other realm. Thus, a speech on the floor of Congress from a whale.

The long "pagan" battle of western poetry against state and church, and the survival of the muse down to modern times, show that in a sense poetry has been a long and not particularly successful defending action. Defending "the groves"—sacred to the Goddess—and logged, so to speak, under orders from Exodus 34:13: "You shall destroy their images and cut down their groves."

The evidence of anthropology is that countless men and women, through history and prehistory, have experienced a deep sense of communion and communication with nature and with specific nonhuman beings. Moreover, they have often experienced this communication with a being they customarily ate. People of goodwill who cannot see a *reasonable* mode of either listening to, or speaking for, nature except by analytical and scientific means must surely learn to take this complex, profound, moving, and in many ways highly appropriate worldview of the yogins, shamans, and ultimately all our ancestors into account. One of the few modes of speech that give us access to that other yogic or shamanistic view (in which all are one and all are many, and the many are all precious) is poetry or song.

[This essay is based on a talk given at the Conference on the Rights of the Nonhuman, Claremont, California, spring 1974. It was first published in The Old Ways *(San Francisco: City Lights, 1977).]*

"Energy Is Eternal Delight"

A young woman at Sir George Williams University in Montreal asked me, "What do you fear most?" I found myself answering "that the diversity and richness of the gene pool will be destroyed," and most people there understood what I meant.

The treasure of life is the richness of stored information in the diverse genes of all living beings. If the human race, following on some set of catastrophes, were to survive at the expense of many plant and animal species, it would be no victory. Diversity provides life with the capacity for a multitude of adaptations and responses to long-range changes on the planet. The possibility remains that at some future time another evolutionary line might carry the development of consciousness to clearer levels than our family of upright primates.

The United States, Europe, the Soviet Union, and Japan have a habit. They are addicted to heavy energy use, great gulps and injections of fossil fuel. As fossil fuel reserves go down, they will take dangerous gambles with the future health of the biosphere (through nuclear power) to keep up their habit.

For several centuries western civilization has had a priapic drive for material accumulation, continual extensions of political and economic power, termed "progress." In the Judeo-Christian worldview humans are seen as working out their ultimate destinies (paradise? perdition?) with planet earth as the stage for the drama—trees and animals mere props, nature a vast supply depot. Fed by fossil fuel, this religio-economic view has become a cancer: uncontrollable growth. It may finally choke itself and drag much else down with it.

The longing for growth is not wrong. The nub of the problem now is how to flip over, as in jujitsu, the magnificent growth energy of modern civilization into a nonacquisitive search for deeper knowledge of self and nature. Self-nature. Mother Nature. If people come to realize that there are many nonmaterial, nondestructive paths of growth—of the highest and most fascinating order—it would help dampen the common fear that a steady state economy would mean deadly stagnation.

I spent a few years, some time back, in and around a training place. It was a school for monks of the Rinzai branch of Zen Buddhism, in Japan. The whole aim of the community was personal and universal liberation. In this quest for spiritual freedom every man marched strictly to the same drum in matters of hours of work and meditation. In the teacher's room one was pushed across sticky barriers into vast new spaces. The training was traditional and had been handed down for centuries—but the insights are forever fresh and new. The beauty, refinement, and truly civilized quality of that life has no match in modern America. It is supported by hand labor in small fields, gathering brushwood to heat the bath, drinking well water, and making barrels of homemade pickles.

The Buddhists teach respect for all life and for wild systems. A human being's life is totally dependent on an interpenetrating network of wild systems. Eugene Odum, in his useful paper "The Strategy of Ecosystem Development," points out how the United States has the characteristics of a young ecosystem. Some American Indian cultures have "mature" characteristics: protection as against production, stability as against growth, quality as against quantity. In Pueblo societies a kind of ultimate democracy is practiced. Plants and animals are also people and, through certain rituals and dances, are given a place and a voice in the political discussions of the humans. They are "represented." "Power to *all* the people" must be the slogan.

On Hopi and Navajo land, at Black Mesa, the industrial world is eating away at the earth in the form of strip-mining. This to provide electricity for Los Angeles. The defense of Black Mesa is being sustained by traditional Indians, young Indian militants, and longhairs. Black Mesa speaks to us through old stories. She is said to be sacred territory. To hear her voice is to give up the European word *America* and accept the new-old name for the continent, "Turtle Island."

The return to marginal farmland on the part of some young people is not some nostalgic replay of the nineteenth century. Here is a generation of white people finally ready to learn from the Elders. How to live on the continent as though our children, and on down, for many ages, will still be here (not on the moon). Loving and protecting this soil, these trees, these wolves. Natives of Turtle Island.

A scaled-down, balanced technology is possible, if cut loose from the cancer of exploitation/heavy industry/perpetual growth. Those who have already sensed these necessities and have begun, whether in the country or the city, to "grow with less" are the only counterculture that counts. Electricity for Los Angeles is not exactly energy. As Blake said, "Energy is eternal delight."

[This was first published in the "Plain Talk" section of Turtle Island *(New York: New Directions, 1974).]*

Earth Day and the War Against the Imagination

On Earth Day 1990, April 22, I was asked to speak at a regional gathering at Bridgeport Crossing on the South Yuba River of northern California. Several hundred country people (most of whom I knew) gathered in a gently sloping meadow just above the brisk sparkly stream. The stage was framed by live oak, rocks, and the canyon stretching up and back to the east. There were booths and eats from the different groups, music, and a stunning "hot-trash" fashion show. Public lands people (the state parks district director, the Bureau of Land Management area manager, and others) were mixed in with the locals and their passions, but everyone was easy, friendly, and willing to talk. A delicate, delicious spring shower came up as we stood there. Everybody was delighted. This is some of what I said.

*I*t's wonderful to be gathered together again here in the lap of the South Yuba River—old friends and new, committed to protecting and enhancing the quality of life in the foothills. This little basin has been appreciated by human beings for at least seven thousand years. The Nisenan people who preceded us lived west along the Yuba down to the junction with the Feather (at what we now call Marysville) and eastward along the ridges up to the winter snowline at around four thousand feet. They had a rich culture, with stories, music, ceremonies, and a deep knowledge of plants and animals. I hope the time will come when we who live in the foothills will start our story with the Nisenan as our previous teachers and spiritual ancestors, rather than with the brief era of the gold miners, the forty-niners, who tediously dominate the local official mythology and decorate our county seal. The miners' lives were bold and ingenious, but we need not limit our narrative simply to the anecdotes of a half century of mineral exploitation—as lively as they were. We're here today to make deeper connections to the earth.

I was a speaker at the first Earth Day twenty years ago. The gathering I attended was at Colorado State University at Greeley. A sky of dark threatening rain clouds and scattered shafts of sunlight came down over the thousands of students sitting on the plaza, as speakers and singers held forth. There were young men with hair in braids, owl feathers woven into them. Practically everyone was wearing shorts and hiking shoes. A slight girl wore a T-shirt that proclaimed, "I am an Enemy of the State." That first Earth Day was not exactly a beginning, but it was a hinge, a turn around a corner. It marked the gradual waning of the need for antiwar activism and the swinging of our energies toward the struggle for the health of the earth. It brought a whole generation of students, and many others who had never much thought of nature before, into a movement in defense of life and death, of the whole process of nature.

There had been an environmental movement before that, of course. It was called the conservation movement, and it dated back to the turn of the century. Its particular concerns were proper public lands management practices, the establishment of wilderness areas, and wildlife survival. The movement had a lot of passion under its somewhat staid exterior. Its activists were often closet pagans and nascent Green radicals, as we can tell from their writings. Even Dave Foreman, one of the founders of Earth First!, says he was a Young Republican in his campus days, and came into radical environmentalism from the public lands conservation movement.

Earth Day 1970 heralded some new developments. The conservation movement became part of something called environmentalism or ecology. From its modest but scrappy beginnings it became one of the most successful democratic movements in the history of American politics. The issues are real, and Congress was moved to write and pass legislation for air, water, and wildlife before it even quite knew what it was getting into. Today the ongoing pressure of environmental issues (plus good scientific critique and sometimes good economics) is pushing the Forest Service and Congress reluctantly to reform forest management practices on public lands. It is helping the cities to turn toward public transportation and recycling. Environmentalists have become players in national and world politics, making them also subject to dilution and appropriation. And more deeply, ecological thought has become a model for an entirely new and different way for human beings to see themselves in the world.

Some of our world-scale issues—to recap familiar ground—are the following:

- World deforestation and soil loss. The rapidly disappearing moist tropical rain forests are the most critical example. Another consists of the huge mining operations launched (without much control) in many Third World countries.

- The rapid loss of biological diversity, the endangerment or extinction of animal and plant species.

- Worldwide water and air pollution. Toxic and radioactive substances have been released into the biosphere with drastic potential results, such as increased cancer rates, sterility, genetic damage, and so on.

- Loss of local cultures, languages, skills, and knowledges, which were developed over millennia. Small communities everywhere are being ground down.

- Loss of heart and soul. This is serious! To lose our life in nature is to lose freshness, diversity, surprise, the Other—with all its tiny lessons and its huge spaces.

These problems are being caused to some degree by combinations of the following:

Worldwide *overpopulation*: let no one say it is not a problem and a cause of problems. The present world population is about 5.5 billion. It was half that in 1950, and one-tenth of that about 1650. It will be 8.5 billion by 2025. There are 1.5 billion people in the Third World who will soon be running out of firewood. The effect of this is further deforestation, on the brushwood level, and further degradation of soils.

Even if economic and social justice were achieved for all people, there would still be a drastic need for ecological justice, which means providing plenty of land and water for the lives of nonhuman beings.

At the same time there are 500 million cars in the developed world. The *unequal distribution of wealth* in the world causes endless social turmoil and intensifies the destruction of nature. The top 20 percent of American households earn more than 43 percent of all income. The same 20 percent hold 67 percent of net worth and nearly 90 percent of net financial assets. The median net worth of the top 1 percent of households is twenty-two times greater than the median of the remaining 99 percent. Median net financial assets of the top 1 percent are 237 times greater than those of the other 99 percent. The richest one-half of 1 percent—about 430,000 households—own 40 percent of corporate stocks. White median net worth is 11.7 times that of blacks. Sixty-seven percent of black households have zero or negative net worth. (These statistics come from an article by David R. Francis in the *Christian Science Monitor*, March 30, 1990.) In this kind of world, people will simply scratch and scrabble with each other to get ahead, and natural resources or endangered species get short shrift.

These dynamics are shaped by, and feed back into, the *industrial economy* and the world economic system. The economic sector is not only a major polluter (and profit taker) but so far it has not accounted for the actual cost of its destructive impact on such "common-pool resources" as air and water. The public and all the rest of nature pick up costs that should be shared by corporate polluters. Some people pay with their health or lives. The economic system of the developed world favors and promotes unsustainable economic growth and has its own reasons (cheap labor, for one) for not taking population issues seriously.

There are socially and politically entrenched *attitudes and institutions* that reinforce our misuse of nature and our cruelty toward each other. Our major civilizations objectify and commodify the natural world. They regard nature as a mere inanimate resource and a target of opportunity. I could say that this is bad metaphysics, but it is worse than that: it is a failure of imagination. Failures of compassion and charity are failures of imagination. As Diane di Prima wrote in a remarkable poem, *Rant*,

> The only war that matters is the war against
> the imagination.

We can easily see some of these problems at work here on the western slopes of the northern Sierra Nevada. One is out-of-control growth. Where will it end? There are some who are making fortunes off it. Such people, or their companies, are often based elsewhere in the state, and they have no loyalty to *any* community or place. It is more complicated for those who take their home seriously and hope that it will be livable and viable for their grandchildren—as some Native Americans say, "up to the seventh generation."

Real estate and business people like to argue that economic growth and development are inevitable. Local boosters might say it's selfish to try to put the brakes on growth. I'd answer most emphatically, *It is not selfish for any community or neighborhood to try to find ways to check unwanted growth and expansion in its own backyard.* Indeed, those who try to shove growth down our throats are precisely the people who profit from it. Are they not permitted to act in their own self-interest? This is a nation in which the pursuit of profit is a totally acceptable goal. Is not the pursuit of *quality* an equally acceptable goal? No apologies are due for trying to hold the line against either disruptive growth or intrusive industrial uses. In doing so we help to sustain community values and the biological viability of our landscape.

The forces of excessive growth and resource exploitation are international, but whatever we accomplish here sends a message elsewhere. There *are* limits; there is such a thing as an optimum carrying capacity for any given region. The estimate of carrying capacity is based on measures of water resources, air quality, adequate space for both human housing and healthy wildlife populations, job-producing resources and their long-range potential, and other biological and health-related criteria. The concept of "carrying capacity" makes human welfare its primary concern but includes the welfare of the whole natural community as well. As an old hunting-gear catalog once put it, "Where wildlife cannot survive, humans cannot survive." In February of this year, the Nevada County Irrigation District announced that the county is at the limit of its water use right now, and further water can only be created by conservation. From a carrying-capacity standpoint I firmly believe that the planet is already overpopulated and that—would it were possible—we should aim at a gradual reduction of numbers—voluntarily, of course—over future decades.

But it *would* be irresponsible if in the process of blocking some unwanted expansion into our own neighborhood, we took no care for the plight of other areas where the same incursion might turn up next. In obstructing something as clearly questionable as, say, toxic waste disposal siting, we would want to be supportive of our friends elsewhere who must fight the very same threat that we may have driven out. The fact that one community could succeed in holding something unwanted at bay should be a source of hope and strength to other communities, and not be dismissed as just a case of "sending it elsewhere." When people everywhere present a united front in their resistance to toxics or bad development projects, then that finally sends a message to the corporations and government agencies: *This is unacceptable. Quit doing it or find a radically different way.* If a community concludes that it doesn't want to be exposed to herbicides and pesticides, or to host nuclear waste locally, or to make way for a huge dam, that is not irresponsible or wrong, it is a people's choice. It might cost somebody some money and disrupt some plans, to be sure. But so does the war on crime. Crime provides millions of jobs, just like pesticides and armaments, but we do make a certain effort to control it.

So we should not waver in political activism. We need to get better at talking to elected officials and government staff people, better at research and critique, better at alliances with other groups, and more savvy about how the state and county people think and work. We should study the maps, read the small print in the environmental impact reports, and wade through the National Forest long-range management plans. We need to keep going to meetings and chambers and readily standing up to speak our piece.

Then let's celebrate the fact that we have a place to be. It may be hot, dry, and rocky here, scary with forest fires every summer; it may not be as flowery as the Alps or as green as Oregon, but why indulge in comparisons? It's our spot on earth—these mountains of northern California, Western Hemisphere, on this great blue-green globe. It is in our power, as people of the place, to shape the future here. Even though it has been logged hard and burned hot, an old-growth forest with a fire ecology can be brought back, and we will be trading a few marvelous knot-free sawlogs to the valley people in about two centuries. The wild herbs, the springs, the boulders with their glyphs will teach us. The rivers will run clearer than before, and even the fisher and the marten will return. Our houses will go on, and the children of the future will find ways to tell finer stories than we, as newcomers, could ever hope to.

So let's keep walking the hills and learning trails, flowers, birds, old cemeteries, old mine shafts, forgotten canyons. Keep on holding potlucks, forest ecology classes for kids, sweat lodges, classes on bark beetles, high-country ski tours, poetry readings, and watershed meetings. We need to stay loose, smart, creative, and wild.

The wild is imagination—so is community—so is a good time. Let's be tough but good-natured Green or Rainbow warriors, make cause with wild nature, and have some ferocious fun while doing it.

[1990]

Nets of Beads, Webs of Cells

ECOSYSTEMS, ORGANISMS, AND
THE FIRST PRECEPT IN BUDDHISM

*T*he primary ethical injunction of Buddhism is known as the First Precept. It is against hurting and taking life, *ahimsa* in Sanskrit, glossed as meaning "cause no unnecessary harm." Not eating flesh is a common consequence of this precept in the Buddhist world, which has largely consisted of agrarian peoples. This has posed a thorny question for normally tolerant Buddhists in the matter of how to regard the spiritual life of people in those societies for whom eating fish or animals may well be a matter of economic necessity. My own home place is beyond the zone of adequate water and good gardening soil, so my family and I have grappled with this question, even as we kept up our lay Buddhist life.

I have plenty of neighbors for whom Buddhism is not even on the map. I know hunters and antihunters, usually decent people on both sides, and have tried to keep my mind open to both. As a student of hunting and gathering cultures, I've tried to get some insight into fundamental human psychology by looking at the millennia of human hunting and gathering experience. I have also killed a few animals, to be sure. On two occasions I put down deer that had been wounded by sport hunters and had wandered in that condition into our part of the forest. When I kept chickens, we maintained the flock, the ecology, and the economy by eating excess young roosters and, at the other end of the life cycle, by stewing an occasional elderly hen. In doing this I experienced one of the necessities of peasant life worldwide. They (and I) could not but run their flock this way, for anything else would be a luxury—that is to say, uneconomic.

Also my hens (unlike commercial hens who are tightly caged) got to run wild and scratch all day, had a big rooster boyfriend, and lived the vivid and sociable life of jungle fowl. They were occasionally taken away by bobcats, raccoons, wild dogs, and coyotes. Did I hate the bobcats and coyotes for this? Sometimes, taking sides with the chickens, I almost did. I even shot a bobcat that had been killing chickens once, a fact of which I am not proud. I probably could have come up with a different solution, and I now think that one must stand humbly aside and let the Great System go through its moves. I did quit keeping chickens, but that was because it was not practical. Happy loose flocks cannot compete with factory egg production, which reduces hens to machines (but protects them from bobcats). (On a deep level I do not think I can approve of the domestication of birds and animals; too much is taken out of their self-sufficient wild natures.)

As for venison, for many years several families in this area have carefully salvaged fresh roadkill deer rather than let flesh go to waste. (But then, letting it feed vultures or carrion beetles is no waste . . .) And by keeping a sharp eye on the roadside I have saved myself the quandary of whether to hunt or not to hunt deer. Fewer and fewer Californians are hunting. But in place of hunters we have a fine resurgent cougar population, and sometimes find their kills in the woods, not far from the house at that.

The public and private forests and grasslands of the western Sierra Nevada make up a sizable ecosystem marked by pines, oak, songbirds and owls, raccoons, deer, and such. The web of relationships in an ecosystem makes one think of the Hua-yen Buddhist image of Indra's net, where, as David Barnhill describes it, "the universe is considered to be a vast web of many-sided and highly polished jewels, each one acting as a multiple mirror. In one sense each jewel is a single entity. But when we look at a jewel, we see nothing but the reflections of other jewels, which themselves are reflections of other jewels, and so on in an endless system of mirroring. Thus in each jewel is the image of the entire net."

This perception of a "sacramentalized ecosystem" lies behind the ceremonies of compassion and gratitude in foraging cultures, where a special respect is paid to the spirits of the game. Wild-plant gathering and gardening also call for respectful attention to the lives of the plants; almost as much mindfulness is asked of the vegetarian as of the hunter.

The very distinction "vegetarian/nonvegetarian" is too simple. Some populations, especially in India and Southeast Asia, are deliberate Buddhist and Hindu vegetarians, but most of the rest of the people of the Third World are semi-vegetarians by default. They are grateful for a little fish or chicken when they can get it. When and where people can live by grains and vegetables alone and get adequate nutrition, it is to be applauded. But there are people of the high latitudes, of the grasslands and deserts and the mountains, who have always relied on much nonplant food. Most people of the world have always had to live by a mixed food economy. Shall Buddhists then consider them beyond the pale? Surely the Bodhisattva spirit does not allow us to reject the other cultures and food economies of the world out of hand. As for modern food production, although it is clear that the beef economy of the developed world is a wasteful luxury, it is doubtful that the Third World could easily get by without cows, chickens, pigs, sheep, and the life of the sea.

Americans, Australians, New Zealanders, and some Europeans are the largest per capita meat consumers of the world. In the developed world vegetarians are usually educated members of a privileged class. Most North American Buddhists have no real need to eat meat, so the choice is theirs. (We then need to study our dependence on fossil fuel agriculture, which produces vegetables and grains in a manner that degrades soil, air, and water and which endangers the health of underpaid immigrant laborers.)

But the real question is how to understand more deeply this First Precept. When Oda Sessō Rōshi, my teacher at the Daitoku-ji monastery, came to the koan in the *Mumonkan*, "Nansen Kills a Cat," he chose not to sit in the high chair but sat on the tatami, on the same level as the *unsui* (monks). He said, "This is a case that can be easily misunderstood, and we in Japan have on some occasions perhaps abused it." At the time I thought he was referring to the apparent lack of resistance on the part of the Zen establishment to the emergence of Japanese militarism in the thirties, leading to World War II. Now I think that he was indicating that *anyone* in a discussion who raises the question of deliberately taking life should be sitting right on the floor. One cannot be too humble about this issue. As I listened to his formal Zen lecture back in 1961, I must confess I felt a certain righteousness, because I had been a lifelong pacifist (and on-and-off vegetarian) and thought I knew how to understand the precept. Not so easy.

I had also noticed that even some of the masters (let alone the monks) ate fish when away from the monastery. One time I was visiting at the temple of a rōshi near Mount Fuji and asked him why it is that some priests and monks eat meat or fish. He responded heartily, "A Zen man should be able to eat dog shit and drink kerosene." My own teacher was a strict vegetarian. But he once said to me, "Just because I eat pure food, and some of the other priests do not, does not mean that I am superior to them. It is my own way of practice. Others have other ways. Each person must take the First Precept as a deep challenge, and find his own way through life with it."

In my natural curiosity I like to know where food has come from and who it was, plant or animal. (Okra is a member of the *Hibiscus* genus, originally from Africa! Tomatoes, tobacco, potatoes, and jimsonweed are all Solanaceae together, with those trumpet-shaped flowers. I love such facts.) My family and I say grace and do a little meditation on our food before meals, just as is done on a larger scale in *sesshin*, "meditation weeks," with the meal verses.

The First Precept goes beyond a concern just for organic life. Yet our stance in regard to food is a daily manifestation of our economics and ecology. Food is the field in which we daily explore our "harming" of the world. Clearly it will not do simply to stop at this point and declare that the world is pain and suffering and that we are all deluded. We are called instead to practice. In the course of our practice we will not transform reality, but we may transform ourselves. Guilt and self-blame are not the fruit of practice, but we might hope that a *larger view* is. The larger view is one that can acknowledge the simultaneous pain and beauty of this complexly interrelated world. This is what the image of Indra's net is for. So far it has been the earlier subsistence cultures of the world, especially the hunters and gatherers, who have—paradoxically—most beautifully expressed their gratitude to the earth and its creatures. As Buddhists we have something yet to learn on that score. Animals and plants live mutually on each other, and throughout nature there is a constant exchange of energy—a cycle of life-and-death affairs. Our type of universe is described in the sutras as a realm of *kama*, of biological desire and need, which drives everything. Everything that breathes is hungry. But not to flee such a world! Join in Indra's net!

None of what I have been saying is to be seen as a rationalization or justification for "breaking" the precept. As Ryo Imamura recently wrote, "in Buddhism there is no such thing as a 'just' war." If we were to find ourselves going against the precept in some drastic situation and killing or injuring someone else in (say) self-defense, we must not try to justify it. We can only say this was my decision, I regret that it happened, and I accept whatever results it may entail.

The precept is the Precept, and it stands as a guide, a measure, an ideal, and a koan. It cannot be a literal rule, as if it were one of the Ten Commandments. "Take no life" or "commit no harm" is impossible to keep perfectly. The Jains of India tried to take *ahimsa* to its literal (not logical) conclusion, and the purest among them started an institution of starving themselves to death as a moral act. But this is violence against one's own body.

Every living thing impinges on every other living thing. Popular Darwinism, with its emphasis on survival of the fittest, has taken this to mean that nature is a cockpit of competitive bloodshed. "Nature red in tooth and claw," as the Europeans are fond of quoting. This view implicitly elevates human beings to a role of moral superiority over the rest of nature. More recently the science of ecology, with its demonstrations of coevolution, symbiosis, mutual aid and support, interrelationship, and interdependence throughout natural systems, has taught us modesty in regard to human specialness. It has also taught us that our understanding of what is and is not "harmful" within the realm of wild nature is so rudimentary that we should not even bother to take sides between predators and prey, between primary green producers and detritus-side fungi or parasites, or even between "life" and "death."

Thich Nhat Hanh once said at a gathering of Buddhist Peace Fellowship leaders at Green Gulch, a Zen farming community in northern California, that we should be grateful for any little appearance of *ahimsa* wherever it is found in this world. I believe he said that if one officer in a battle leads his troops with a bit more spirit of *ahimsa* than another, it is to be appreciated. It is my sense of it, then, that we must each find our own personal way to practice this precept, within quite a latitude of possibilities, understanding that there will be no complete purity and in any case not indulging in self-righteousness. It is truly our "existential koan." This is why I have glossed it, in the Mahayana spirit, as "commit no unnecessary harm."

One can wonder what the practice of *ahimsa* is like for the bobcat, in the bobcat Buddha realm. As Dōgen says, "dragons see water as a palace," and for bobcats, the forest is perhaps an elegant *jikido*, dining hall, in which they murmur *gathas* of quiet appreciation to quail, sharing them (in mind) with demons and hungry ghosts. "You who study with Buddhas should not be limited to human views when you are studying water" (*Mountains and Waters Sutra*). And what world is it for quail? I only know this: at death, my death and suffering are my own, and I hope I will not blame my distress on the tiger (or cancer, or whatever) that has brought me down. Of the tiger I would simply hope to ask, "Please, no waste." And maybe growl along with her.

There is an old Zen story of a teacher finding a single discarded chopstick on the drain. He scolds the dishwasher monk, saying, "You have taken the life of this chopstick." This story is used to illustrate how deeply the First Precept reaches. We can look thus at a wasted chopstick and understand how it has been harmed. But then it should also be added, "You might even be killing a rain forest," as the use of disposable wood chopsticks—in staggering quantities—in Japan and America suggests.

Did the master know that next step? Probably not. Buddhist compassion for creatures sometimes meant purchasing and then ceremonially releasing caged pigeons and captured fish, the focus being on individual creatures. Individual lives are only part of the story. Even as the Buddhists were practicing vegetarianism and kindness to creatures, wild nature in China suffered significant species extinction and wholesale deforestation between the fifth and the fifteenth centuries A.D. India too was vastly deforested well before modern times. Now, with insights from the ecological sciences, we know that we must think on a scale of a whole watershed, a natural system, a habitat. To save the life of a single parrot or monkey is truly admirable. But unless the forest is saved, they will all die.

The whole planet groans under the massive disregard of the precept of *ahimsa* by the highly organized societies and corporate economies of the world. Thousands of species of animals, and tens of thousands of species of plants, may become extinct in the next century. To nourish living beings we must not be content simply to have a virtuous diet. To save all beings, we must work tirelessly to maintain the integrity of these mandala-like places of habitat, and the people, creatures, and Buddhas who dwell in their palace-like spaces.

[This essay was published in 1990, in an earlier version, as part of an exchange with David Barnhill and others in Ten Directions, *the journal of the Los Angeles Zen Center, under the title "Indra's Net as Our Own." My thanks to David Barnhill for pushing this discussion along.]*

A Village Council
of All Beings

COMMUNITY, PLACE, AND THE
AWAKENING OF COMPASSION

*In Leh, capital of the district of Ladakh in northern India, the
Ladakh Ecological Development Group and its founder, Helena
Norberg-Hodge, held a conference called "Rethinking Progress" in September 1992. Speakers came from Europe, India, and
the United States; ecologist Stephanie Mills and I were among
them. I had been asked to speak on "ecology and spirituality."
There were some eminent monks in attendance and a remarkable lay Buddhist philosopher, Tashi Rapges. I felt honored to be
given the opportunity to enlarge on my thoughts about how ecological insights and bioregional organizing might help a Buddhist society under siege from the witless expansionism of the
industrial world.*

Oecology, as it used to be spelled, is the scientific study of relationships, energy transfers, mutualities, connections, and cause-and-effect networks within natural systems. By virtue of its findings, it has become a discipline that informs the world about the danger of the breakdown of the biological world. In a way, it is to Euro-American global economic development as anthropology used to be to colonialism—that is to say, a kind of counterscience generated by the abuses of the development culture (and capable of being misused by unscrupulous science mercenaries in the service of the development culture). The word *ecological* has also come to be used to mean something like "environmentally conscious."

The scientist, we are told, seeks to be objective. Objectivity is a semisubjective affair, and although one would aspire to see with the distant and detached eye of a pure observer, when looking at natural systems the observer is not only affecting the system, he or she is inevitably part of it. The biological world and its ecological interactions *are* this world, our very own world. Thus, ecology (with its root meaning of "household science") is very close to economics, with its root meaning of "household management." Human beings, biology and ecology tell us, are located completely within the sphere of nature. Social organization, language, cultural practices, and other features that we take to be distinguishing characteristics of the human species are also within the larger sphere of nature.

To thus locate the human species as being so completely within "nature" is an unsettling step in terms of the long traditions of Euro-American thought. Darwin proposed evolutionary and genetic kinship with other species. This is an idea that has been accepted intellectually but not personally and emotionally by most people. Social Darwinism flourished for a while as a popular ideology justifying nineteenth-century imperialism and capitalism, with an admiring emphasis on competition. The science of ecology corrects that emphasis and goes a step further. It acknowledges the competitive side of the process, but also brings forward the coevolutionary, cooperative side of interactions in living systems. Ecological science shows us that nature is not just an assembly of separate species all competing with each other for survival (an urban interpretation of the world?) but that the organic world is made up of many communities of diverse beings in which the species all play different but essential roles. It could be seen as a village model of the world.

An ecosystem is a kind of mandala in which there are multiple relations that are all-powerful and instructive. Each figure in the mandala—a little mouse or bird (or little god or demon figure)—has an important position and a role to play. Although ecosystems can be described as hierarchical in terms of energy flow, from the standpoint of the whole all of its members are equal.

But we must not sentimentalize this. A key transaction in natural systems is energy exchange, which includes the food chains and the food webs, and this means that many living beings live by eating other beings. Our bodies—or the energy they represent—are thus continually being passed around. We are all guests at the feast, and we are also the meal! All of biological nature can be seen as an enormous *puja*, a ceremony of offering and sharing.

The intimate perception of interconnection, frailty, inevitable impermanence, and pain (and the continuity of grand process and its ultimate emptiness) is an experience that awakens the heart of compassion. It is the insight of bodhicitta that Shantideva wrote so eloquently about. It is the simultaneous awakening of a personal aspiration for enlightenment and a profound concern for others.

Ecological science clearly throws considerable light on the fundamental questions of who we are, how we exist, and where we belong. It suggests a leap into a larger sense of self and family. It seems clear enough that a consequence of our human interdependence should be a social ethic of mutual respect and a commitment to solving conflict as peacefully as possible. As we know, history tells a different story. Nonetheless, we must forge on to ask the next question: how do we encourage and develop an ethic that goes beyond intra-human obligations and includes nonhuman nature? The last two hundred years of scientific and social materialism, with some exceptions, have declared our universe to be without soul and without value except as given value by human activities. The ideology of development is solidly founded on this assumption. Although there is a tentative effort among Christians and Jews of goodwill to enlarge their sense of ethics to include nature (and there have been a few conferences on "eco-Christianity"), the mainstream of Euro-American spirituality is decidedly human-centered.

Asian thought systems (although not ideal) serve the natural world a little better. Chinese Taoism, the Sanātana ("eternal") Dharma of India, and the Buddhadharma of much of the rest of Asia all see humanity as part of nature. All living creatures are equal actors in the divine drama of awakening. As Tashi Rapges said, the spontaneous awakening of compassion for others instantly starts one on the path of ecological ethics, as well as on the path toward enlightenment. They are not two.

In our contemporary world, an ethic of concern for the nonhuman arrives not a moment too soon. The biological health of the planet is in trouble. Many larger animals are in danger of becoming extinct, and whole ecosystems with their *lakhs* (one hundred thousands) of little living creatures are being eliminated. Scientific ecology, in witness to this, has brought forth the crisis discipline of conservation biology, with its focus on preserving biodiversity. Biodiversity issues now, as never before, bring local people, industries, and governments into direct and passionate dialogue over issues involving fisheries, marine mammals, large rare vertebrates, obscure species of owls, the building of huge dams or road systems, and more.

The awakening of the Mind of Compassion is a universally known human experience and is not created by "Buddhism" or any other particular tradition. It is an immediate experience of great impact, and Christians, Jews, Muslims, communists, and capitalists will often arrive at it directly—in spite of the silence of their own religions or teachings on such matters. The experience may often be completely without obvious ethical content, a moment of leaving the hard ego self behind while just seeing, just being, at one with some other.

Much of India and the Far East subscribes in theory at least to the basic precept of nonharming. *Ahimsa*, nonviolence, harmlessness, is described as meaning "cause the least possible harm in every situation." Even as we acknowledge the basic truth that every one of us lives by causing some harm, we can consciously amend our behavior to reduce the amount of practical damage we might do, without being drawn into needless feelings of guilt.

Keeping nature and culture healthy in this complicated world calls us to a kind of political and social activism. We must study the ways to influence public policy. In the Western Hemisphere, we have some large and well-organized national and international environmental organizations. They do needed work, but are inevitably living close to the centers of power where they lobby politicians and negotiate with corporations. In consequence they do not always understand and sympathize with the situations of local people, village economies, tribal territories, or impoverished wageworkers. Many scientists and environmental workers lose track of that heart of compassion and their memory of wild nature.

The actualization of the spiritual and political implications of ecology—ensuring that it is more than rhetoric or ideas—must occur place by place. Nature happens, culture happens, somewhere. This grounding is the source of bioregional community politics. Joanna Macy and John Seed have worked with the image of a vague sort of "Council of All Beings." The idea of a *"Village* Council of All Beings" suggests that we can get place-specific. Imagine a village that includes its trees and birds, its sheep, goats, cows, and yaks, and the wild animals of the high pastures (ibex, argali, antelope, wild yak) as members of the community. The village councils, then, would in some sense give all these creatures voice. They would provide space for all. Any of the Ladakhi and Tibetan village territories in this part of the world certainly should include the distant communal pastures (*p'u*) and the subwatersheds as well as the cultivated fields and households. In the case of Ladakh and indeed all of India, when a village is dealing with government or corporation representatives, it should insist that the "locally used territory" embrace the whole local watershed. Otherwise, as we have too often seen, the government agencies or business interests manage to co-opt the local hinterland as private or "national" property and relentlessly develop it according to an industrial model.

We also need an education for the young people that gives them pride in their culture and their place, even while showing them the way into modern information pathways and introducing them to the complicated dynamics of world markets. They must become well informed about the workings of governments, banking, and economics—those despised but essential mysteries. We need an education that places them firmly within biology but that also gives them a picture of human cultural affairs and accomplishments over the millennia. (There is scarcely a tribal or village culture that doesn't have some sort of music, drama, craft, and story that it can be proud of.) We must further a spiritual education that helps children appreciate the full interconnectedness of life and encourages a biologically informed ethic of nonharming.

All of us can be as placed and grounded as a willow tree

along the stream—and also as free and fluid in the life of the whole planet as the water in the water cycle that passes through all forms and positions roughly every two million years. Our finite bodies and inevitable membership in cultures and regions must be taken as a valuable and positive condition of existence. Mind is fluid, nature is porous, and both biologically and culturally we are always fully part of the whole. This ancient nation of Ladakh has always had such people living in it. Some of the beautiful young men and women here today will master the modern world, keep up the Dharma, and continue to be true people of Ladakh.

[Upper Indus River watershed,
western Tibetan plateau, town of Leh,
nation of Ladakh,
province of Jammu and Kashmir,
state of India,
September 7–11, 1992]

II

Aesthetics

Goddess of Mountains
and Rivers

FOREWORD TO EDWARD SCHAFER'S
THE DIVINE WOMAN:
DRAGON LADIES AND RAIN MAIDENS
IN T'ANG LITERATURE

*I*n the belly of the furnace of creativity is a sexual fire; the flames twine about each other in fear and delight. The same sort of coiling, at a cooler, slower pace, is what the life of this planet looks like. The enormous spirals of typhoons, the twists and turns of mountain ranges and gorges, the waves and the deep ocean currents—a dragonlike writhing.

Western civilization has learned much in recent years of its archaic matrifocal roots. Part of that has been the recovery of a deeper sense of what "muse" means, and a new understanding of the male-female play in our own hearts. Robert Graves's poetic essay, *The White Goddess*, has been pivotal in disclosing the continuity of a muse-magic tradition. The poet-muse relationship is usually seen from the male side only, for we live in cultures, both East and West, that have been dominated by men for several thousand years. Of all males through those patriarchal years, the poets and artists were most apt to go beyond the one-sided masculine ethos and draw power from that other place, which the Chinese would call the yin side of things. It is likely that men become creative when they touch the woman in themselves, and women become creative when they touch the woman in the man in themselves.

We work here with the faint facts of a Neolithic past and the actual facts of a planetwide interconnected web of living beings. The totality of this biosphere is called by some Gaia, after the ancient Greek earth goddess. It should be no surprise that one singer will be inspired by the heavy breast of a slender girl and another by the wind whipping through a col plastering the cliffs with gleaming rain.

When English-speaking readers first came onto Chinese poetry in translation, about sixty years ago, there was a sigh of relief. It was refreshing to get away from romanticism and symbolism and to step into the cool world of Chinese lyric poetry. Here were poems of friendships and journeys, moments of tender thought for wives and children, praise of quiet cottages. There were also the slightly more passionate poems for courtesans and concubines, a minor mode. We were ignorant of the fact that the poems are far more complex and formal in the original than any translation could let us know. Chinese poetry in translation helped us find a way toward a clear secular poetic statement, and the melancholy tone of the T'ang is echoed in the nature elegies of some poets today.

But where are the women? We see nature, but where is the muse? This book answers that question. The calm male lyric strain in China conceals a wilder thread that goes back to prehistoric times. The Chinese perceived mountains and rivers as numinous; special bends in the rivers or the contorted strata of high-piercing pinnacles were seen as spots of greater concentration of *ch'i*, spirit power. Even the rational Confucianists of the T'ang era believed that nature was alive, believed in fox-ladies and ghosts. From earliest times, the "yin"—shady side, moist, fertile, and receptive—was identified as "female." The "yang"—sunny side, fertilizing, warming, dry—as "male." And it is written, the yin and yang together make the Tao. The fifth-century B.C. *Tao Te Ching* is full of the echo of a great goddess: spirit of the valley, mother of the ten thousand things, marvelous emptiness before being and nonbeing. The dance of yin-yang energies in nature (mist on the mountain peaks, rainbows and rain squalls, rocky cliffs and swirling streams, tumbling flight of flocks of birds) becomes the image vocabulary of Chinese erotic poetry.

So we can trace, from the far Chinese past into literate bureaucratic times, the continuing presence of semihuman goddesses of mountains and streams. They are not meaty, all breasts and hips like the goddesses of India, or athletic like those of Greece. The difference is that the Chinese have not projected a human physical image onto the world but have seen the natural world draw close sometimes and temporarily assume a faint human shape. The goddess of the Wu mountain, the "Divine Woman," is first seen as a glimmering figure of cloud, mist, and light. In the poets' description she has an exquisitely high-class gauzy garb and is heartbreakingly remote. The mountain's name, Wu, means "female shaman." Such women were very powerful in the Neolithic and Bronze ages. They survive among the people right down to the present.

The literary evidence for those beginnings comes from the *Ch'u Tz'u*, "Songs of the South," a collection from the old southern kingdom of Ch'u. Some of these poems are spirit possession songs in which both men and women call out for unearthly lovers. These texts are the starting point for the shamanistic thread in Chinese poetry.

As upper-class Chinese culture becomes increasingly male-oriented through history, this strain grows weak and precious. Li Ho is the sole exception. By the T'ang dynasty, the focus of Edward Schafer's study, the lore of the river goddesses and the Divine Woman is a story of disillusion and inconclusiveness.

The eighth-century poet Li Ho is a poet of the wilder muses. He seeks them not only in the wilderness but also among the singsong girls of the pleasure barges, floating with lantern and zither in the night. These women were beautifully refined, consciously embodying the archetype, and too expensive for him. They modeled themselves on the archaic images of cloud, rainbow, and river. Rainbow: a trope in Chinese poetry for woman's ecstasy.

The Mountain Goddess of the "Nine Songs," part of the *Ch'u Tz'u*, is described as dressed

In a coat of fig leaves with a rabbit-floss girdle . . .
Driving tawny leopards, leading the striped lynxes;
Her cloak of stone orchids, her belt of asarum:
She gathers sweet scents to give to one she loves.

(Translated by David Hawkes)

(Twice on rocky mountain pinnacles, once at seventeen in the Cascades of Washington and again at forty-two in the Daisetsu range of Hokkaido, sitting in a cloud with no view at all, settling back on the boulders in the mist, I found myself inexplicably singing. That song was for her.)

Chinese male culture is profoundly ambiguous about nature and women. The best poets were often failed bureaucrats. Their submissive loyal wives were like the cultivated fields, the singsong girls like the wilderness. That explains why, finally, in the Chinese prose tales, the goddess becomes evil. What had earlier been seen as a fulfilling surrender to the spirits of nature became a fear of being pitilessly drained by the ungovernable wild. The goddess image had turned lethal.

In *Woman Warrior*, Maxine Hong Kingston quotes the folk sayings she heard as a child about daughters: "Girls are maggots in the rice," "It is more profitable to raise geese than daughters," "Feeding girls is feeding cowbirds." Being goddesses in civilized times never did real women much good. Kingston wants to become the woman warrior Hua Mu Lan.

I knock my forehead three times on the ground to Edward Schafer for writing this book. I think again of rain maidens, and remember the water cycle.

The Water Cycle:
The 1.5 billion cubic kilometers of water on the earth are
split by photosynthesis and reconstituted by respiration
once every two million years or so.

    Scientific American. The Biosphere *(San Francisco: 1970)*

So she moves through us. The work of art has always been to demonstrate and celebrate the interconnectedness: not to make everything "one" but to make the "many" authentic, to help illuminate it all. Besides being an elegantly written delight, *The Divine Woman* illuminates a lesser-known side of Chinese poetic tradition, and is another step toward getting the energies back in balance.

[Edward Schafer (who died in 1991) was one of my teachers at the University of California at Berkeley, where I studied oriental languages in the early fifties. He was elegant, crusty, demanding, precise, witty, and a true lover of poetry (as well as of gemstones, incense, birds . . .). We kept up a correspondence after I moved to Japan, and he told me that he had written his (remarkable) Golden Peaches of Samarkand *as a sort of handbook for poets. I started reading all of his scholarly works in this light.* The Divine Woman: Dragon Ladies and Rain Maidens in T'ang Literature *(San Francisco: North Point Press, 1980), another sort of guide for poets, partly came out of Dr. Schafer's later research into medieval Taoism.]*

What Poetry Did in China

Chinese poetry, at its finest, seems to have found a center within the tripod of humanity, spirit, and nature. With strategies of apparent simplicity and understatement, it moves from awe before history to—a deep breath before nature. Twentieth-century English-language translations make this poetry into "plain tone and direct statement," and in this form Chinese poetry has had a strong effect on occidental poets tired of heroics and theologies. That this actually elaborate and complex poetic tradition should have made such a contribution to occidental modernism is rather curious. Yet it can be understood as having something to do with the twentieth-century thirst for naturalistic secular clarity. Chinese poetry provided the exhilarating realization that such clarity can be accomplished in the mode of poetry.

The introduction to the fifth-century B.C. Chinese "classic of poetry," the *Shih Ching*, says, "Poetry is to regulate the married couple, establish the principle of filial piety, intensify human relationships, elevate civilization, and improve public morals." This is to suggest, reasonably enough, that poetry in a halfway-functioning society has an integrative role. We do recognize that poetry can make one remember one's parents, celebrate friendships, and feel tender toward lovers. Poems give soul to history and help express the gratitude we might sometimes feel for the work and sacrifices of our predecessors. Poetry strengthens the community and honors the life of the spirit.

Chinese poetry in the era of the *Shih Ching* had no obvious sensibility for landscapes and large-scale nature. So what's missing in this early appraisal of poetry from the "straight" side of Chinese culture—the world of the early literati—is an idea of how poetry might give human beings a window into the nonhuman. We know that the arts lend us eyes and ears that are other than human, pointing toward other biologies, other realms. From the fourth to the fourteenth centuries, the poetry of China reached far (but selectively) into the world of nature. Contemporary occidental poetry has been influenced by that aspect, too.

But now at the end of the twentieth century most societies are not even halfway functioning. What does poetry do then? For at least a century and half, the socially engaged writers of the developed world have taken their role to be one of resistance and subversion. Poetry can disclose the misuse of language by holders of power, it can attack dangerous archetypes employed to oppress, and it can expose the flimsiness of shabby made-up mythologies. It can savagely ridicule pomp and pretension, and it can offer—in ways both obvious and subtle—more elegant, tastier, lovelier, deeper, more ecstatic, and far more intelligent words and images.

Poetry also serves as a mode of speaking for our dreams and for the deep archetypes. Poetry will not only integrate and stabilize, it will break open ways out of the accustomed habits of perception and allow one to slip into different possibilities—some wise, some perhaps bizarre, but all of them equally real, and some holding a promise of further new angles of insight. The Chinese, with their Confucian focus on the family and the community, and the Chinese Buddhists with their return-to-the-world-to-help vows would argue that one will bring back and share whatever one has found. Some poems are, in this way, truly a voice from outside.

This mix of elegant pragmatism and shamanistic vision is what seems to have marked the poetry of the people who lived in the basins of the Chiang and the Ho (the Yangtze and the Yellow) rivers over the last two millennia.

[This essay, written in 1994, was drawn from talks given over the years.]

Amazing Grace

*T*here are two basic modes of learning: "direct experience" and "hearsay." Nowadays most that we know comes through hearsay—through books, teachers, and television—keyed to only a minimal ground of direct contact with the world. (The "world" is perceived as a rolling outdoor space with weather above, obstructions underfoot, and plants, people, animals, buildings, and machines occupying various niches.)

Hearsay is the great organizer of this apparent chaos via myth, science, or philosophy. Not too long ago there were no writing systems, and the worldview/myth/frameworks came to young listeners as long stories chanted in the evening. These old stories are the foundation stones of what the Occident calls classics, and indeed all literature.

In a completely preliterate society the oral tradition is not memorized but *remembered*. Thus, every telling is fresh and new, as the teller's mind's eye re-views the imagery of origins or journeys or loves or hunts. Themes and formulas are repeated as part of an ever-changing tapestry composed of

both the familiar and the novel. Direct experience, generation by generation, feeds back into the tale told. Part of that
direct experience is the group context itself, a circle of listeners who murmur the burden back or voice approval, or
snore. Meaning flashes from mind to mind, and young eyes
sparkle.

All later civilized educations are by degrees removed
from this primacy of together-hearing. An urban cosmopolitanism is gained, with the loss of a keen sense of the integra-
tion of human and natural systems. In the tales of the Ainu,
who were the indigenous people of Japan and who still live
in the north, gods and animals speak in the first person as
well as human beings, and the several worlds of sense experience and imagination are knit together.

The many motifs of oral literature found worldwide,
which at least prove that humanity enjoys the same themes
over and over, are not heard as part of some comparative
study demonstrating the unity of humanity, but as out of the
minds, hills, and rivers of the place—maybe through the
mouth of a bear or salmon. A people and a place become one.

Such were the Ainu, on one level a remnant population of
a few bands, isolated for centuries from the "centers of world
civilization." But it's all here in their tales: the planetwide
themes; the great adventures of love, sorcery, and battle
("The Epic of Kotan Utunnai"); and the almost uniquely
Ainu telling of tales direct from nonhuman entities, a mode
of "interspecies communication."

On another level the Ainu are at the center of an archaic
internationalism. Their big island was a meeting place of
circumpolar hunting culture pathways with Pacific seacoast

cultures. In the practices they lived by are some of the purest teachings of those old ways that survive: the sacramental food-chain mutual-sharing consciousness that was likely the basic religious view of the whole Northern Hemisphere Paleolithic. This view, after a lapse of many millennia, clearly has relevance to us again: the planet earth—Gaia—must now be seen as *one system*.

The people of precivilized times or places knew their specific watershed ecosystems and mastered those details with beautiful and empirical precision. Natural systems, even in small areas, are of the utmost complexity, and to be understood must be grasped in their wholeness. This means, so to speak, leaving the trail and walking uphill and down, through the brush. The trail is what village people use as a straight line between garden plot and garden plot. Hence, "linear." The forest, for hunting and gathering people, must be grasped, visualized as a field: "Where do you suppose the deer are moving today?" The Ainu term *iworu*, "field of force," means simply biome, or territory, but has spirit-world implications as well.

So an Ainu group would live along a river in a house facing east, fire at the center. The upstream field was a forest, swamp, and mountain wilderness, penetrated by the trails of the hunters. (The arctic brown bear of Hokkaido is as large as a grizzly.) The downstream field was the coast and the ocean, full of herring and salmon, cod and crab, and before the Japanese came, rich in seals, sea lions, and whales. When men returned from hunting and fishing, and women from gathering plants for food, fiber, medicine, poison, and dyes, they sat by the fire. Men would carve intricate bas-relief designs on knife sheath and quiver. Women wove, sewed, and embroidered the graceful linear swooping designs that are instantly recognizable as Ainu. An elder perhaps told these stories. The life of mountains and rivers flowed from their group experience, through speech and hands, into a fabric of artifacts and tales that was a total expression of their world and of themselves. As the Ainu saw it, from the inner mountains upstream and from the sea depths downstream, game came as visitors. Master of the one realm is Bear, master of the other is Killer Whale. The deer or salmon would leave behind their flesh bodies in exchange for being entertained with songs, stories, and wine by the humans. Humans are good musicians, as the whole world knows. Returning to their sea or mountain home with gifts, the animal or fish spirits would hold another party in the spirit realm, and many would agree it was good to visit the human world, and more would soon go. Thus, cycles in and out of a real landscape, and cycles in and out of life and death, attended by that highest of pastimes—singing and feasting (in or out of mask) with food and friends.

Paradoxically, only now, in the last years of the twentieth century, can this view be understood for its real worth. Millennia of rapacious states spilling out of their boundaries to plunder the resources and people within reach created a false image of limitless space and wealth on the planet, available for whoever had the weapons, organization, and willingness to kill without saying thanks. Through no wisdom of its own, but out of necessity, industrial civilization in particular is now forced to realize that there are limits, and that there is a life-support system composed of millions of subsystems all working or playing together with amazing grace.

Through Donald Philippi's translations, the Ainu suggest to us with great clarity that this life-support system is not just a mutual food factory, it is mysteriously *beautiful*. It is what we are. We now see the Ainu not as a fading remnant, but as elders and teachers whose playful sense of their own bioregion points a way to see and live on our planet as a whole.

[This essay was originally published as the foreword to Songs of Gods, Songs of Humans: The Epic Tradition of the Ainu, *translated by Donald Philippi (San Francisco: North Point Press, 1982).]*

The Old Masters and the Old Women

FOREWORD TO SŌIKU SHIGEMATSU'S
A ZEN FOREST

*T*he Mojave Indians of the lower Colorado River put all the energy they gave to aesthetic and religious affairs into the recitation of long poetic narratives. Some of the epics are remarkably precise in describing the details of the vast basin and range deserts of the Southwest, but the raconteurs held that they were all learned in dreams. By another sort of inversion, the world of Ch'an (Zen) Buddhism with its "no dependence on words and letters"—and unadorned halls, plain altars, dark robes—created a large and very specialized literary culture. It registers the difficulty of the play between verbal and nonverbal in the methods of the training halls. The highly literate Zen people were also well acquainted with secular literature, and they borrowed useful turns of phrase from any source at all, to be part of the tool kit, to be employed when necessary, and often in a somewhat different way. Then Ch'an texts, Chinese poems, Buddhist sutras, Taoist and Confucian classics, and proverbial lore were sifted one more time. This was done in Japan in the sixteenth and seventeenth centuries, and the result was the *Zenrin Kushū*, "Phrases from the Zen Forest." The greater part of the phrases gathered come from Chinese poetry, so that R. H. Blyth could say that the *Zenrin Kushū* is "the Zen view of the world on its way through poetry to haiku."

A Zen Forest is not quite like any collection of quotations or selections from "great literature" that has been seen before. Eichō Zenji, who did the basic editing, and his successors obviously knew what they were looking for. Sōiku Shigematsu's introduction tells about that.

But the *Zenrin Kushū* selections could not have the terse power and vividness they do were it not for the richness of the parent material. First, the terseness. It's all from Chinese. (Readings given in the appendix of the book, to accompany the Chinese characters, are in a form of literary Sino-Japanese and do not represent the pronunciation of the Chinese or the word order. They are read in this way by Japanese Zen students.) The Chinese language is mostly monosyllabic, with word-order grammar, and can be very economical. There is a long-established culturewide delight in sayings and quotes, and there is a special lore of ambiguity and obscurity that plays on the many homonyms in the language. Early books such as the *I Ching* and the Taoist essays abound in "dark sayings." The Zen phrase anthologies do not draw on deliberately obscure sayings, tongue twisters, traditional riddles, and the like. With the exception of quotes taken from the texts of their own school, they present us with selections from the public body of sayings and quotes. Poems are never quoted whole, so that in this case the obscurity (especially for the Western reader) comes from the absence of context. When the Zen phrase is actually an old proverb, such as

> To sell dog meat,
> displaying a sheep's head

several levels of meaning are instantly clear. In Chinese this would literally be "Hang sheep head sell dog meat." Another proverb that comes into the phrase book is

> One who flees
> fifty steps
> Sneers at the other
> who's done a hundred.

English is a relatively parsimonious language, but the Chinese for this is literally "Fifty steps sneer him ahead one hundred." The context is running away from battle.

The most numerous type of Zen phrase consists of couplets borrowed from poems of five characters to a line. This section is called the "paired fives," and there are 578 such couplets in the Baiyō Shoin edition of the *Zenrin Kushū*. (Mr. Shigematsu has done away with the traditional arrangement of Zen phrases by number of characters. His original and personal sequencing makes the book perhaps easier to read straight through.) Seven-character lines, both single and paired, make up the next largest body of quotes, also from poetry.

Chinese poetry takes the crisp virtue of the language and intensifies it a turn again. It is also the one area of the literature where personal sentiment—vulnerability, love, loneliness—is to be found in an otherwise dry and proper terrain. The very first teaching of Buddhism that Chinese intellectuals took to heart, in the fourth and fifth centuries A.D., was that of impermanence. It fit well with the political experience of the times, the troubled Six Dynasties. The lyric poetry of the era was also full of woe and gloom. So almost from the beginning *shih* poetry has had a line to Buddhism. The Chinese (and almost everyone else) consider the T'ang poetry of the eighth century to be the crown of their literature. The poems of this period, infinitely superior to the weepy Six Dynasties lyrics, are the ones most often raided for Zen quotes. T'ao Ch'ien is a notable exception. We are speaking especially of the poets Wang Wei, Li Po, Tu Fu, Han Shan, and Liu Tsung-yüan. Although some were Buddhists, this does not matter to a Zen phrase. The power of image and metaphor—the magic of poetry—not ideology, is what counts. Contemporary with these poets are the great creative Ch'an masters Shen-hui, Nan-yüeh, Ma-tsu, Pai-chang, and Shih-tou. For whatever reason, the Golden Age of Chinese poetry is also the Golden Age of Ch'an. Twelfth-century Ch'an masters who gathered and edited koan books out of the anecdotes and lives of T'ang masters were also reading and quoting the T'ang poets.

Many of the poems from which the Zen editors plucked quotes have been widely known by almost all Chinese and educated Japanese for centuries now. Some of them have entered the territory of *su hua*, or "common sayings." Such is Tu Fu's

> The country is ruined: yet
>   mountains and rivers remain.
> It's spring in the walled town,
>   the grass growing wild.

The context here is the destruction of the capital during the An Lu-shan rebellion. Tu Fu was not a Buddhist, yet his way of being and working came close to the essence. Burton Watson says of Tu Fu, "Tu Fu worked to broaden the definition of poetry by demonstrating that no subject, if properly handled, need be unpoetic. . . . There is evidence to suggest that he was versed in the lore of herbs and medicinal plants, and perhaps this knowledge gave him a special appreciation for the humbler forms of natural life. Some of his poems display a compassion for birds, fish, or insects that would almost seem to be Buddhist inspired. Whatever the reason, he appears to have possessed an acute sensitivity to the small motions and creatures of nature. . . . Somewhere in all the ceaseless and seemingly insignificant activities of the natural world, he keeps implying truth is to be found."

The poets and the Ch'an masters were in a sense just the

tip of the wave of a deep Chinese sensibility, an attitude toward life and nature that rose and flowed from the seventh to the fourteenth centuries and then slowly waned. The major Ch'an literary productions, *Wu-men Kuan, Ts'ung-jung Lu, Pi-yen Lu, Hsü-t'ang Lu*, are from the twelfth and thirteenth centuries. It was a second Golden Age of Ch'an and another era of marvelous poetry, one in which many poets were truly influenced by Ch'an. The most highly regarded Sung dynasty poet, Su Shih (Su Tung-p'o), was known as a Ch'an adept as well as poet and administrator.

> Valley sounds:
> the eloquent
> tongue—
> Mountain form:
> isn't it
> Pure Body?

This is part of a poem by Su Shih. The Japanese master Dōgen was so taken with this poem that he used it as the basis for an essay, *Keisei Sanshoku*, "Valley sounds, mountain form." Sung-dynasty Ch'an had a training system that took anecdotes and themes from its own history and lore and assigned them as subjects of meditation. The tradition that emphasized this, the Rinzai sect, is also called the Zen that "looks at sayings." The complementary school called Sōtō, which cut down on the use of old sayings, is also called "silent-illumination Zen." They were both brought from China to Japan on the eve of the Mongol invasions. Japan inherited and added on to its own already highly developed sense of nature the worldview of T'ang and Sung.

Robert Aitken Rōshi has described koans (and by implication the "phrases from the Zen forest") as "the folklore of Zen." Borrowed in part from the folklore of a whole people, their use as Zen folklore is highly focused. These bits of poems are not simply bandied about among Zen students as some kind of in-group wisdom or slangy shorthand for larger meanings. They are used sparingly, in interviews with the teacher, as a mode of reaching even deeper than a "personal" answer to a problem: as a way of confirming that one has touched base with a larger Mind. They are not valued for the literary metaphor but for the challenge presented by the exercise of translating the metaphor into the life of the body, into insight and action. They help the student bring symbols and abstractions back to earth. Zen exquisitely develops this possibility—yet it's not far from the natural work of poems and proverbs anyway. Someone has said proverbs are proverbs because they are so true.

So if Zen has koans for folklore, the world has folklore for koans. Proverbs and short poems the world around are of like intensity and suggest equal depths. Though Mr. Shigematsu has chosen to eliminate the Zen phrases of less than four words from his collection, it helps to know how they work, and why. What would be the power of a one-word Zen phrase? I think of Harry Roberts's account of Yurok Indian upbringing: if one did something foolish, all that Elder Uncle had to say was

Well!

and a youngster would go off to ponder for hours.

Let us lock eyebrows with the seventh-century B.C. Greek
poet Archilochus, a mercenary soldier:

>—So thick the confusion
>Even the cowards were brave

>—The crow was so ravished by pleasure
>That the kingfisher on a rock nearby
>Shook its feathers and flew away

>—Into the jug
>Through a straw
>>*(Translated by Guy Davenport)*

A Bantu riddle:

>A black garden
>With white corn—
>>the sky and stars.

And the Philippines:

>The houseowner was caught;
>the house escaped
>>through the window.
>>>—Fishnet.

And the Koyukon of the Alaskan Yukon:

>—Flying upward
>ringing bells in silence:
>the butterfly

>—Far away, a
>fire flaring up:
>red fox tail

> —We come upstream
>   in red canoes:
>   the salmon.
>   *(Translated by R. Dauenhauer)*

And the Samoans:

> —When the old hen scratches,
>   the chicks eat beetles.

The Hawaiians:

> —Not all knowledge
>   is contained in your dancing school.

And finally the people of Kentucky:

> "My feet are cold" one says,
>   and the legless man replies:
>   "So are mine.
>   So are mine."
>   *(Wendell Berry)*

But even beyond the fascinating Ch'an/Zen and world folklore implications of this collection, it stands on its own as a kind of "poem of poems." We can read Mr. Shigematsu's excellent translations and follow his creative sequencing with an availability that has been earned for us by the modernist poetry of this century. Hugh Kenner speaks of "our renewed pleasure in the laconic and the expletive functions of language" in his introduction to Archilochus. Let this book be read then for the enjoyment of the far-darting mind, and skip for the time any notions of self-improvement. It is a new poem in English, winnowed out of three thousand years of Chinese culture, by some of the best minds of the East. It's also the meeting place of the highest and the most humble: the great poets and the "old women's sayings," as proverbs are called. Arthur Smith, speaking of mandarin officials of nineteenth-century China, said they were "well known to spice their conferences and their conversation with quotations from 'the old women' as naturally as they cite the Four Books [of Confucius]."

For this book to exist, the Ch'an masters of the past, the poets of the twentieth century, and the old women must have joined hands.

[This essay was originally the foreword to Sōiku Shigematsu's book of translations from the Zenrin Kushū, *entitled* A Zen Forest: Sayings of the Masters *(Tokyo: Weatherhill, 1981). Mr. Shigematsu is a priest of the Rinzai school of Zen Buddhism, with a temple in Shizuoka prefecture.]*

A Single Breath

*I*n this world of onrushing events, the act of meditation—
even just a "one-breath" meditation: straightening the back,
clearing the mind for a moment— is like a refreshing island
in the stream. Although the term *meditation* has mystical
and religious connotations for many people, it is a simple
and plain phenomenon. Deliberate stillness and silence. As
anyone who has done this knows, the quieted mind has
many paths, many of them tedious and ordinary—and then
sometimes unexpected. But meditation is always instructive,
and there is ample testimony that a practice of meditation
pursued over months and years brings some degree of self-
understanding, serenity, focus, and self-confidence to the
person who stays with it.

The making of poems and the traditions of deliberate attention to consciousness are both as old as humankind. Meditation looks inward, poetry holds forth. One is for oneself,
the other is for the world. One enters the moment, the other
shares it. But in practice it is never entirely clear which is

doing which. In any case, we do know that in spite of the contemporary public perception of "poetry" and "meditation" as special, exotic, and difficult, they are both as old and as common as grass. The one begins with people sitting still and reflecting, and the other with people making up songs and stories and performing them.

People often confuse meditation with prayer, devotion, or vision. They are not the same. Meditation as a practice does not address itself to a deity or present itself as an opportunity for revelation. This is not to say that people who are meditating do not occasionally think they have received a revelation or experienced visions. They do. But to those for whom meditation is their central practice, a vision or a revelation is seen as just another phenomenon of consciousness and as such is not to be taken too seriously. The meditation practicer is simply experiencing the ground of consciousness, and in doing so avoids excluding or excessively elevating any thought or feeling. To do that one must release all sense of the "I" as experiencer, even the "I" who might think it is privileged to communicate with the divine. It is in sensitive areas such as these that a teacher can be a great help. This is mostly a description of the Buddhist tradition, which has hewed consistently to a nondevotional, nontheistic practice over the centuries.

Poetry has also been part of Buddhism from early on. There has, however, been some ambivalence toward the arts in Buddhism. The Chinese Ch'an masters often said, "The lowest class of monk is the one who indulges in literature." (We have to remember that blame is often praise in the Ch'an world.) Some of the finest poets of China were acknowledged Ch'an adepts—such as Po Chü-i and Su Tung-p'o, to name just two.

The Ch'an training halls, with their unconventional Dharma discourses and vivid pantomimed exchanges, and the Chinese lyric poems, *shih*, with their lucid and allusive brevity, were shaping each other. Ch'an teachers and students have always written poems. In formal koan practice a student is often called upon to present a few lines of poetry from the Chinese canon as a proof of the completeness of his or her understanding of the koan, a practice called *jukugo*, or "capping verses," in Japan. These exchanges have been described in *A Zen Forest* by Sōiku Shigematsu, who has handily translated hundreds of the couplets borrowed from Chinese poetry and proverb. They are intense:

> Words, words, words—fluttering drizzle and snow.
> Silence, silence, silence—a roaring thunderbolt.

> Bring back the dead!
> Kill the living!

> This tune, another tune—no one understands.
> Rain has passed, leaving the pond brimming in the
>   autumn light.

> The fire of catastrophe   has burned out all
> Millions of miles no mist, not a grain of dust!

> One phrase after another
> Each moment refreshing.

Every poetic tradition has forms and specialties of its own. There is a formal craft of poetry as much as there are institutionalized religions. They are not always separate. The Buddhist world has produced numerous poets and singers of the Dharma whose works are admired and loved. Mila Repa, whose songs are still known by heart among Tibetans, and Bashō, whose haiku are read worldwide, are perhaps the most famous. Yet poetry (and the literary world) has sometimes been perceived as dangerous to the spiritual career. We can appreciate Ikkyu's probing poem titled "Ridiculing Literature":

> Humans are endowed with
> > the stupidity of horses and cattle.
> Poetry was originally a
> > work out of hell.
> Self-pride, false pride,
> > suffering from the passions,
> We must sigh for those taking this path
> > to intimacy with demons.

Ikkyu, a fifteenth-century Japanese Zen master and a fine (and strikingly original) poet himself, laughingly ridiculed his fellow poets, knowing as he did the distractions and temptations that might come with literary aspirations. His "intimacy with demons" is not to be interpreted in the light of the occidental romance with alienation, however. In Japanese art, demons are funny little guys, as solid as horses and cows, who gnash their fangs and cross their eyes. Poetry is a way of celebrating the actuality of a nondual universe in all its facets; its risk is that it declines to exclude demons. Buddhism offers demons a hand and then tries to teach them to sit. But there are tricky little poetry/ego demons that do come along, tempting us with "suffering" or with "insight," with success or failure. There are demons practicing meditation and writing poetry in the same room with the rest of us, and we are all indeed intimate. This didn't really trouble Ikkyu.

For a gathering of poetry yogins and yoginis at Green Gulch Zen Center one spring, I wrote, "We have to appreciate the Mind that floats our many selves, gives shelter to our hard won information and word hoards, and yet remains a sea of surprises. (Whatever made people think Mind isn't rocks, fences, clouds, or houses? Dōgen's ingenuous question.) Meditation is the problematic art of deliberately staying open as the myriad things experience themselves. Another one of the ways that phenomena 'experience themselves' is in poetry. Poetry steers between nonverbal states of mind and the intricacies of our gift of language (a wild system born with us). When I practice zazen, poetry never occurs to me, I just do zazen. Yet one cannot deny the connection."

On seeing Ikkyu's poem (and my comments), my friend Doc wrote me from his fish camp:

Ikkyu says, "Humans are endowed with the stupidity
    of horses and cattle."
I think Ikkyu is full of shit.
Humans are endowed with a stupidity all their own.
Horses and cattle know what to do.
They do it well.
He is right about poetry as a work out of hell.

We ought to know.

Phenomena experience themselves as themselves.

They don't need poetry.

We are looking at a mystery here.

How do these things have such an obstinacy

and yet are dependent on my consciousness?

When I practice fishing with two teenagers

poetry never occurs to me.

But later it does.

I can go over the whole day.

Hooray! That's what being human is all about.

It is just as much a weakness as a strength.

You say language is (a wild system born with us).

I agree.

It is wilder than wild.

If we were just wild we wouldn't need language.

Maybe we are beyond wild.

That makes me feel better.

Kanaka Creek.
Doc Dachtler.

Beyond wild. This can indeed include language. Poetry is how language experiences itself. It's not that "the deepest spiritual insights cannot be expressed in words" (they can, in fact) but that "*words* cannot be expressed in words." So our poems are full of *real presences*. "Save a ghost," you might be asked by your teacher, or an owl, or a rain forest (or a demon). Walking that through and then putting a poem to it are steps on the way toward realization. But the path has many switchbacks, and a spiritual journey is strewn with almost as many land mines as a poet's path. Let us all be careful (and loose as a goose) together.

Following the practice of meditation must have a little to do with getting "beyond wild" in language. Spending quality time with your own mind is humbling and, like travel, broadening. You find that there's no one in charge, and are reminded that no thought lasts for long. The "marks" of Buddhist teaching are impermanence, no-self, the inevitability of suffering, connectedness, emptiness, the vastness of mind, and a way to realization. A poem, like a life, is a brief presentation, a uniqueness in the oneness, a complete expression, and a gift. In the No play *Bashō* ("Banana Plant"), it is said that "all poetry and art are offerings to the Buddha." These Buddhist ideas interacting with the Chinese sense of poetry are part of the weave that produced an elegant plainness, which we name the Zen aesthetic.

The idea of a poetry of minimal surface texture, with the complexities hidden at the bottom of the pool, under the bank, a dark old lurking, no fancy flavor, is ancient. It is what is "haunting" in the best of Scottish-English ballads, and is at the heart of the Chinese *shih* (lyric) aesthetic. Tu Fu said, "The ideas of a poet should be noble and simple." In Zen circles it is said, "Unformed people delight in the gaudy and in novelty. Cooked people delight in the ordinary." This plainness, this ordinary actuality is what Buddhists call "thusness," or *tathata*.

All is actuality. There is nothing special about *actuality* because it is all right here. There's no need to call attention to it, to bring it up vividly and display it. Therefore the ultimate subject matter of poetry is profoundly ordinary. The really fine poems are maybe the invisible ones, that show no special insight, no remarkable beauty. (No one has ever really achieved writing great poems that have perfectly no insight and beauty—it is only a distant ideal.)

But there will never be just one sort of identifiable "meditation poetry," and gaudiness and novelty are also fully real. There will never be—I hope—one final and exclusive style of Buddhism. We can be frank: in poetry and in meditation you must be shameless, have no secrets from yourself, be constantly alert, make no judgment of wise and foolish, high or low class, and give everything its full due.

Then there might be poems that have the eye that sees the moment and can play freely with what's given:

Teasing the demonic
Wrestling the wrathful
Laughing with the lustful
Seducing the shy
Wiping dirty noses and sewing torn shirts
Sending philosophers home to their wives in time for
    dinner
Dousing bureaucrats in rivers
Taking mothers mountain climbing
Eating the ordinary

We can appreciate that this can be done very quietly. Even plainly. In this complex and elegant theater of samsara.

[This essay was first published as the foreword to Kent Johnson and Craig Paulenich, eds., Beneath a Single Moon: Buddhism in Contemporary American Poetry *(Boston: Shambhala, 1991).]*

Energy from the Moon

What is all this about—the moon and flowers? Saigyō the monk-poet traveled wide and far. His poems speak of monkeys, owls, fishermen, boulders, emptiness, love, war, the Buddhadharma, and old acquaintances of Kyoto days; but most seem to keep coming back to the moon, watched from hundreds of places, hundreds of nights. And Saigyō keeps returning in poem and in body to the flowering cherry, *sakura*, especially the trees that bloom on the hills of Yoshino.

Some of the power in these poems—in Japanese and in the translations offered here—comes from the way the web of attention is spun out on a complex syntax, drawing the mind as far as it will stretch, to be deftly completed with a final phrase that sometimes illuminates all that went before. The effort is meditative, and though the poems look short they go a long way. Haiku, as a shorter form developed later, is a confession that *waka*, Saigyō's form, is difficult to sustain. Haiku, with fewer syllables, moves more quickly; it has even more immediacy, less "mind" in the way. Yet Saigyō's *waka* are as necessary to haiku, and to the mind, as sutras are to koans, as organic evolution is to the cricket of the moment.

So, Saigyō's fifty-year-long Buddhist practice was gazing at moon and flowers, while being in many places. To anyone who has moved on similar paths, it is clear that Saigyō had in fact entered deeply into nature in an experiential way. He was not writing—as some have suggested—merely to a faddish mode. No temple-bound monk could write of climbing a cliff clutching wild azaleas, or of wading a deep river and feeling "washed clean to the base of the heart." The richness of his knowledge of watersheds, seasonal cycles, and organisms brings one back again to the question, why all those moons and flowers?

Moonlight has been pouring down on this planet for billions of years, and for all us beings tuned to—nay, fed by—the energy of the sun, the light of the moon has long been an odd and unsettling force. We have it inside us, like seawater and calcium, a cool light that is always in motion, but stable and recurrent in its changes. The full moon has long been a Buddhist symbol of the Tathagata—of perfect and complete enlightenment. Even enlightenment may be understood as just a fine high phase in a cycle—who does not also love the sight of the new crescent moon in a lavender sunset sky? But Saigyō takes the moon into his mind and out again; it is an opening into another way of seeing this universe in all its space and with its beautiful fragile little creatures.

Though Saigyō wouldn't have thought much about it, the flowers of flowering plants coevolved with insects and are beautiful and sweet-scented for them, not for human beings. The flowering cherries of Japan—*Prunus serrulata* and relatives—produce negligible fruit and are closer to a wild type, the mountain *sakura*, that blooms on the dark conifer hillsides like gleaming clouds. There are a trembling, expectant several days of openness—waiting for the seed to move around—and then they blow off and away. Saigyō says, as if he himself were a bee drawn to the flower, that the masses of blossoms on the slopes of the Yoshino mountains draw him to the depths of the hills for knowledge:

> Yoshino mountains—
> The one who will get to know
> You inside out is I,
> For I've gotten used to going
> Into your depths for blossoms.

The blossoms then are also a way into the inner depths, as well as the more commonly taken symbol of evanescence and youthful beauty.

Some friends and I once hiked through the Yoshino hills at just that time of spring—cherries are still there, pouring down from the heights "like a cascade of white cloud." We went on, up the old mountain yogin's trail, and traversed the whole sacred ridge of Ōmine for four days. Many Japanese readers will know that above and behind Yoshino is the oldest center of Mountain Buddhist (*yamabushi*) practice in the country, a landscape mandala in which Ōmine Peak is the center of the Diamond Realm.

A sense of fleeting life and tiny size in vast calm void on the arm of the curve of the planet makes us also bugs in a realm of flowers. No judgments are made, but through those gazing meditations lies a compassionate, broad view. The moments of loneliness and vacillation ("a warbler lost in a cloud!") are not only human but correct. We and nature are companions, and although authoritative voices do not speak from clouds, a vast, subtle music surrounds us, accessible via clarity and serenity.

All that Saigyō left behind were his poems. There is a modest Japanese restaurant in Sacramento, California, called Kagetsu, "Flowers and Moon." It would be called something else, and there might be no cherry trees planted in Washington, D.C., if Saigyō hadn't lived. There is also the terribly overcrowded, polluted, successful, and confused modern Japan.

Saigyō's poems are masterful mind-language challenges. Bill LaFleur's deeply understanding translations present us with the snakelike energy of the syntax, and the illuminated world that was called out by one man's lifetime of walking and meditating is again right here.

[*This essay appeared as the foreword to* Mirror for the Moon, *William LaFleur's translations of Saigyō's poetry (New York: New Directions, 1978). Saigyō (1118–1190) was a Buddhist monk from the warrior class who lived simply and wandered widely.*]

Walked into Existence

*N*anao Sakaki's poems and presence are known from Tokyo to Amsterdam, New York to London, Maine to San Francisco. He also lives and works, completely at home, in the mountains back of Taos, in the deserts of the lower Rio Grande, in the pine forests of the Sierra Nevada, the subtropical islands of the Ryukyu archipelago, the chilly spruce woods of Hokkaido, the narrow alleys of Kyoto, and the maze of ten thousand bars in Shinjuku, Tokyo. He is one of the first truly cosmopolitan poets to emerge from Japan, but the sources of his thought and inspiration are older than East or West. And newer.

Nanao, "seventh son," was born in a poor village near Kagoshima, the southernmost city of size on Kyushu, which is the most southerly of the main Japanese islands. He was the youngest child in a large family connected with the weaving industry. He was drafted into the air force in World War II, and stationed on the west coast of Kyushu as a radar analyst. His wide reading habits (they discovered a copy of Kropotkin's *Mutual Aid* in his locker) and his far-ranging critical conversations got him into near-serious trouble during the war years, but he scraped through. He sat in on the farewell parties for young kamikaze pilots leaving at dawn the next day for their death, and identified the B-29 that was on its way to bomb Nagasaki on his radar screen. Upon the announcement of the surrender of Japan, his outfit's senior officer ordered the men to prepare to commit mass suicide. Someone luckily turned on a radio to hear the Emperor himself command, in almost incomprehensibly archaic Japanese, that there was no need for soldiers to kill themselves.

The young Japanese men released from the military were almost pariahs in a ruined landscape. The fresh-faced GI occupiers had the democracy and the chocolate, but there was no GI Bill for the ragged boys of a defeated army. Many of them became ardent Marxists, and then ultimately transformed themselves into the world's sharpest businessmen. Some, like Soko Morinaga Rōshi, applied in fear and trembling at the gate of a Zen monastery, to survive years of short sleep and wretched food and become a greathearted teacher in the Buddhist tradition. Nanao Sakaki became a wandering scholar and itinerant artist, and, like Thoreau, the unofficial examiner of the mountains and rivers of all Japan. For fifteen years he walked up and down the land, penetrating backcountry fastnesses and the laborers' ghettos of Osaka. He read widely, in English and European languages and in classical Chinese. At one period he did sculpture, leaving his monumental works in remote localities. His early experimental poems were published in the little magazines of the lively Shinjuku community of semioutlaw intellectuals.

Many of the artists and writers of this broad group became the political and cultural leaders of a small but potent movement, one that is still working for cultural and ethnic diversity, for the preservation of remaining wildlands and wildlife, sustainable agriculture, and a general lowering of the heat within "Japan, Inc."

I first heard Nanao's name in 1962 while traveling to Sri Lanka third class on the old French Messageries Maritime passenger freighter the *Cambodge*, from an Australian writer and roustabout named Neale Hunter. He had met Nanao in Shinjuku and done some preliminary translations of the "Bellyfulls" poems. A summer or two later, in Kyoto, Nanao turned up, and our conversations on the banks of the Kamo River led to a long personal friendship and ongoing collaboration in the art of, not "street theater," but what we might call "fields and mountains theater." Allen Ginsberg was in Kyoto visiting at the time, so ours became a transpacific friendship. We were able to tell Nanao what was happening with poets in Europe and America, and these stories were reciprocated by Nanao's warm and detailed accounts of the amazing number of quietly working, passionate, impoverished, proudly independent writers and thinkers he had met across Japan in his wanderings.

In the late sixties Nanao founded a loose agricultural community on Suwanose Island in the East China Sea. In spite of its remoteness (it was visited by a small ship once every ten days), the "Banyan ashram" had a number of visitors from India, America, and Europe over the years. The young Japanese who settled there from urban worlds have now become fully accounted village members, and are indeed some of the bearers of the endangered arts and crafts of the rapidly fading "Ten Islands" culture.

Nanao first traveled to the United States in 1969, and immediately began to explore the mountains and deserts of the West. His usual Japanese traveling attire of those days—hacked-off shorts, hiking boots, rucksack—was far more socially acceptable in the American West than in Japan. In far deeper ways, Nanao found the landscape, the working people, and the dug-in vernacular poets and activists hospitable. He walked through some of the wildest areas of North America, and began to write some of the poems found in this collection, of place, vast space, plants and their migrations, spiritual enlightenment and vernacular ecstasy. The subtropical East China Sea carpenter and spearfisherman found himself equally at home in the desert, so much so that on one occasion when an eminent traditional Buddhist priest boasted to Nanao of his lineage, Nanao responded, "I need no lineage; I am desert rat."

Over the last decade Nanao has traveled internationally. Based in Japan, he has visited Europe, the People's Republic of China, Australia, and come several more times to the U.S. In 1986 he was in Japan, engaged in actively resisting a large airport slated to be built atop a pristine coral reef on one of the southern Okinawan islands, and reading his poems and presenting his several plays with local groups. At least two volumes of his poems are published in Japan now, where he is being recognized as a unique and powerful voice. His spirit, craft, and knowledge of history make him—whether he likes it or not—an exemplar of a lineage that goes back to the liveliest of Taoists, Chuang-tzu. His poems were not written by hand or head, but with the feet. These poems have been sat into existence, walking into existence, to be left here as traces of a life lived for living—not for intellect or culture. And so the intellect is deep, the culture profound. And this kind of intellect and culture is precisely what the people of China and Japan have appreciated as the real meaning of learning and culture for millennia. For all his independence Nanao Sakaki carries the karma of Chuang-tzu, Hsieh Ling-yun, Lin-chi, En-no-gyoja, Saigyō, Ikkyu, Bashō, Ryokan, and Issa in his bindle. This is the gift that he hands over to the twenty-first century, with the Grand Canyon and penguins rolled in!

But Nanao's work is also truly unique. I know of no poems with quite this slant: compassionate, funny, deceptively simple, cosmic, deeply radical, free. Most of them were written in Japanese and later translated into English by Nanao with the aid of friends. A few were written first in English and then translated back to Japanese. He has translated some American poems into Japanese, and the language he has used is notable for its rigor, its almost Chinese precision and austerity, which is perhaps Nanao's response to the crispness of the English tongue.

And I'll also say he has bony knees, a dark tanned face, odd toes, a fine chanting voice, a huge capacity for spirits, a taste for top-quality green tea, and three beautiful children. His work or play in the world is to pull out nails, free seized nuts, break loose the rusted, open up the shutters. You can put these poems in your shoes and walk a thousand miles!

[This essay was published as the foreword to Nanao Sakaki's Break the Mirror *(San Francisco: North Point Press, 1987).]*

The Politics of Ethnopoetics

*T*his politics is fundamentally concerned with the question of what occidental and industrial technological civilization is doing to the earth. The earth (I'm just going to remind us of a few facts) is fifty-seven million square miles, 3.7 billion human beings, evolved over the last four million years; plus two million species of insects, one million species of plants, twenty thousand species of fish, and 8,700 species of birds; constructed out of ninety-seven naturally occurring surface elements with the power of the annual solar income of the sun. That is a lot of diversity.

Yesterday David Antin told how the Tragedians asked Plato to let them put on some tragedies. Plato said, "Very interesting, gentlemen, but I must tell you something. We have prepared here the greatest tragedy of all. It is called The State."

From a very early age I found myself standing in awe before the natural world. I felt gratitude, wonder, and a sense of protection, especially as I began to see the hills being bulldozed for roads, and the forests of the Pacific Northwest magically float away on logging trucks. I grew up in a rural family in the state of Washington. My grandfather was a homesteader in the Pacific Northwest. The economic base of the whole region was logging. In trying to grasp the dynamics of what was happening, rural state of Washington, 1930s, Depression, white boy out in the country, German on one side, Scotch-Irish-English on the other side, radical—that is to say, sort of grass-roots union, IWW—and socialist parents, I found nothing in their orientation (critical as it was of American politics and economics) that could give me an understanding of what was happening. I had to find that through reading and imagination, which led me into a variety of politics: Marxist, anarchist, and onward.

Now I would like to think of the possibility of a new humanities. Humanities, remember, are a post-Renaissance way of looking at the question of how to shake humans loose from the theological vision of the Middle Ages. But I can't think about our situation in anything less than a forty-thousand-year timescale. That's not very long. If we wanted to talk about hominid evolution we'd have to work with something like four million years. Forty thousand years is a useful working timescale because we can be sure that through the whole of that period humans have been in the same body and in the same mind that they are now. All the evidence we have indicates that imagination, intuition, intellect, wit, decision, speed, and skill were fully developed forty thousand years ago. In fact, it may be that we were a little smarter forty thousand years ago since brain size has somewhat declined on the average from that high point of Cro-

Magnon. It is significant that even the average size of the Neanderthal skull indicates larger brain size than that of the modern human. We don't know why brain size declined. It probably has something to do with "society," if you want to blame it on something. Society has provided buffers and protection of an increasingly complicated order so that as it has become larger in scope, and populations larger in size, it has protected individuals from those demands for speed, skill, knowledge, and intelligence that were common in the Upper Paleolithic. The personal direct contact with the natural world required of hunters and gatherers—men and women both—generated continual alertness.

An appreciation for this archaic and long-maintained intelligence and alertness would have to be part of the foundation of a new humanities. This humanities would take the whole long *Homo sapiens* experience into account, and eventually make an effort to include our nonhuman kin. It would transform itself into a posthuman humanism, which would defend endangered cultures and species alike.

Today we are witnessing an unparalleled waterfall of destruction of the diversity of human cultures, plant species, animal species—of the richness of the biosphere and the millions of years of organic evolution that have gone into it. Ethnopoetics, the study of the poetries and poetics of nonliterate peoples, is like some field of zoology that is studying disappearing species. We must have a concern with this because the cultures that compose and perform such poems and songs are rapidly disappearing.

The major part of the human being's interesting career has been spent as a hunter and gatherer, in "primary" cultures. About twelve thousand years ago, agriculture began to play a small part in some corners of the world. It's only in the last three millennia that agriculture has really penetrated widely. Civilization represents a very small part of human experience—literacy representing an even tinier part, since it's only been in the last two centuries that any sizable proportion of any civilized country has had much literacy. Thus, oral literature—the ballad, the folktale, myth, the songs (the subject matter of "ethnopoetics")—has been the major literary experience of humanity. When we understand that, it becomes all the more poignant that this richness is being swept away.

In the first issue of *Alcheringa*, Jerome Rothenberg and Dennis Tedlock made a statement of intention that I'd like to refer back to, because it seems to me that gathering here, almost five years later, we have the chance to look at those original stated intentions again and see how we've worked with them. Eight points in this statement: "As the first magazine of the world's tribal poetries, *Alcheringa* will not be a scholarly journal of 'ethnopoetics' so much as a place where tribal poetry can appear in English translation and can act (in the oldest and newest of poetic traditions) to change men's minds and lives." Note that, "to change men's minds and lives." "While its sources will be different from other poetry magazines it will be aiming at the struggling and revelatory presentation that has been common to our avant-gardes. Along the way we hope: (1) by exploring the full range of poetics to enlarge our understanding of what a poem may be; (2) to provide a ground for experiments in the translations of tribal/oral poetry and a forum to discuss the possibilities and problems of translation from widely divergent cultures; (3) to encourage poets to participate actively in the translation of tribal/oral poetry; (4) to encourage ethnologists and linguists to do work increasingly ignored by academic publications in their fields, namely to present the tribal poetries as of value in themselves, rather than as ethnographic data; (5) to be a vanguard for the initiation of cooperative projects along these lines between poets, ethnologists, songmen, and others; (6) to return to complex/ 'primitive' systems of poetry, as (intermedia) performance, etc., and to explore ways of presenting these in translation; (7) to emphasize by example and commentary the relevance of tribal poetry to where we are today; (8) to combat cultural genocide in all of its manifestations."

I think that most of us understand what has happened in regard to those areas of interaction described in points two through seven over the last four or five years, so I'm going to concentrate my comments on the two points "combat cultural genocide" and "what a poem may be."

To combat cultural genocide one needs a critique of civilization itself and some thought about what happens when "crossing barriers" takes place—when different, small, relatively self-sufficient cultures begin to contact each other and that interaction becomes stepped up by a historical process of growing populations, growing accumulation of surplus wealth, and so forth. It's probably true that there's a certain basic cross-cultural distrust in small societies that is resolvable through trade, exchange, or periodic gambling games, festivities, and singing together. The sheer fact of distance alone, physical distance between two groups, between two households, makes one group think of those other people as "the others."

The real arms race starts with bronze weapons and certainly with iron. Raiding cultures emerge; that is the first turbulent kind of interface. Some people quit farming and hunting and take up raiding for a living. This goes on today, in what Ray Dasmann calls the relationship between ecosystem cultures and biosphere cultures. Ecosystem cultures are those whose economic base of support is a natural region, a watershed, a plant zone, a natural territory within which they have to make their whole living. Living within the terms of an ecosystem, out of self-interest if nothing else, you are careful. You don't destroy the soils, you don't kill all the game, you don't log it off and let the water wash the soil away. Biosphere cultures are the cultures that begin with early civilization and the centralized state; they are cultures that spread their economic support system out far enough that they can afford to wreck one ecosystem and keep moving on. Well, that's Rome, that's Babylon. It's just a big enough spread that you can begin to be irresponsible about certain specific local territories. Such cultures lead us to imperialist civilization with capitalism and institutionalized economic growth.

The energy we operate by fundamentally is the annual solar income, via agrarian or natural hunting and gathering modes of receiving it, plus your labor—man-for-man, woman-for-woman, labor. Slavery becomes the first concentrated energy hit to speed the expansion of civilized economies. The next big concentrated energy hit is fossil fuels—fossil fuels from the 1880s, responsible for the explosion of all growth curves and consumption curves we see in the world today.

Within that context, we have a number of educated human beings, especially of the occidental world, who, parallel with the worldwide spread of occidental trading habits, became students of other peoples, and (without involving ourselves at this point much with the argument of whether or not anthropology is always imperialism) we can't help but see this as a politically related factor. Anthropological curiosity only occurs if you are a member of an expanding civilization. The contrast to that is to be in a cultural situation where you will not have any particular interest in what other people's cultural habits are, but simply, hopefully, you will respect them. In Zen Buddhism they say, "Mise mono ja nai," which means this is not something we show to people. No radio interviews, no tapings, no videos, no movies, no visitors are permitted in Zen training establishments. It's not for show. It's open to everyone who wishes to participate but it's not for show. That is the sense that insiders have as members of their own culture. They see people who come to them wanting to study (but not participate) as strangely floating around on the surface. Thus, we can begin to imagine how weird our anthropological efforts must look to people who are in that other kind of culture that is ecosystem-based and deeply rooted in its own identity while not doubting in the least the humanity of other human beings.

So now I'd like to tackle this thing about "combat cultural genocide." How do we combat cultural genocide? Has *Alcheringa* combated cultural genocide in the last five years? Have any of us in any focused way combated cultural genocide? Where is cultural genocide taking place? Let's take Brazil. In a recent issue of *Critical Anthropology*, Dr. Jack Stauder makes suggestions to fellow teachers about how to take certain simple academic steps in the right direction. He says if you're going to be an anthropology teacher, you should also be able to teach your students the dynamics of their own culture, at least in the critical area of understanding imperialism and capitalism. If you can't communicate that to your students, then you've got no business talking to them about the Xingu people of the Amazon. He says an anthropologist should be able to teach members of an oppressed culture the dynamics of imperialism and useful economic understanding, insofar as they want to learn it. I know people who don't want to put their heads into those contemporary categories, but if they want to learn it they should be helped. Simply understanding how things work can make the difference between being victimized and being the master of the situation. Dr. Stauder suggests that an anthropologist should play an active political role in society. And that we should ally ourselves to people's struggles everywhere.

Now, Brazil is only one case in point, but it is a very instructive one. People are of course oppressed everywhere, and the destruction of small traditions is taking place in countries of all degrees of complexity. The Brazilian case is particularly touching because it's probably there that the last primary human beings in the world live: a few small groups, apparently, that have not yet even been contacted by expanding civilization. Two hundred and fifty known tribes existed in Brazil in 1900; eighty-seven have become extinct. Between 1900 and 1957 the Indian populations in Brazil dropped from over one million to less than two hundred thousand persons. The population of Brazilian Indians in the Amazon basin is now estimated at less than fifty thousand. The Nambiquara, Cintas Largas, Kadiweu, Bororo, Waura are all examples of threatened populations. This destruction is backed by large multinational corporations; the second largest investor in Brazil is Volkswagen. Volkswagen apparently does not want to convert all its Western Hemisphere profits back into Eurodollars, so it's heavily invested in the development of cattle range in the Brazilian jungle, causing the destruction of forests and replacement of them by grasses to feed the affluent taste for beef of the people of North America. Another is Georgia Pacific, in timber, a company that is also deforesting some of the finest remaining virgin tropical forests of the Philippines on contracts with the Philippine government. Rio Tinto Zinc; Litton Industries doing aerial surveys and mapping; Caterpillar Tractor in vast contracts for pushing out the jungle, going directly across the Xingu park. The Brazilian official statement is "we think the only way for the Indians to improve their health, education, and begin self-development is through development." Now, before you laugh, ask yourself this question: Do you have a good answer to that argument? Do you want to take the position that the Indians of Brazil should be placed in a national park with a fence around it and have absolutely no contact with the civilized world at all? I know as a student of anthropology in the 1950s I became convinced (following the lines of what my teachers were saying) that the traditional cultures of the world were doomed. We could study them, we could try to preserve what we could find of their languages, customs, myths, folktales, ethnobotanic knowledge, and so forth, but it would be quixotic to think that we should invest any political effort in the actual defense of their cultural integrity. The assumption was that there was an almost automatic melting-pot process of assimilation (that was probably OK) under way, and what we had to look for at the other end of the tunnel was a hopeful, international, one-world, humane modernism, fueled by liberal and Marxist ideas. But Marxists, granted the precision of their critique on many points, have a hard time thinking clearly about primitive cultures, and their usual tendency is to assume that they should become civilized. Right? Why do you say that they should not be developed? You want to keep them from having aspirin? Is that even possible?

These strange contradictions. In Argentina there's a national park. The Mapuche of Curruhuinca live there. The forest huts are deteriorating, not owing to laziness but because the park service decrees that no wood may be cut or gathered by the Indians. Surrounded by forest yet disallowed wood and fined if they should dare to cut any. The government provides bundled firewood, but never enough.

The following are quotations from Argentina, but I have heard similar things said in Montana, Utah, Nevada, New Mexico, Arizona, northern California, central Oregon, and so forth. Talking about the Mapuche, a colonel of German origin said, "Are you going to write about them? They're alcoholics and they sleep with their own daughters." A store owner of Arab origin: "But don't worry for them. I hope they die. You had better concern yourself that there will be a good road built." A restaurant owner: "I don't understand them. They starve but they are also so proud that they don't want to become dishwashers." A lawyer with a tourist agency: "The Curruhuincas live marvelously without any shortages at all; by God, you and I would wish we had the same." A high official of Parque Nacional: "What do you want to say about prohibiting their goats? What we want is to throw them out of here. They are lazy, have bad customs, and are dirty. What a spectacle for the tourists. We are studying a project of displacement to another part of the region where they can live as they wish without problems." The official didn't mention that any other region in Neuquen province is desert, bleak and barren, and besides, the Curruhuinca Mapuche belong (and this is acknowledged under Argentinian law) in the area of Lake Lacar, where they are now.

The destruction of cultural diversity goes hand in hand with ecological destruction. I noticed some comments earlier in this conference that seemed to me at least to imply that some of the speakers were in favor of a kind of one-world assimilation of languages and cultures or assumed that some kind of internationalization was a desirable process. The ecological critique goes like this (I quote from Roy Rappaport, "Flow of Energy in an Agricultural Society"): "It may not be improper to characterize as ecological imperialism the elaboration of a world organization that is centered in industrial society and degrades the ecosystems of the agrarian societies it absorbs. The increasing scope of world organization and the increasing industrialization and energy consumption on which it depends have been taken by Western man to virtually define social evolution and progress. What we have called progress or social evolution may be maladaptive. We may ask if the chances for human survival might not be enhanced by reversing the modern trend of successions in order to increase the diversity and stability of local, national, and regional ecosystems even, if need be, at the expense of the complexity and interdependence of international worldwide organizations. It seems to me that the trend toward decreasing ecosystem complexity and stability, rather than threats of pollution, overpopulation, or even energy famine, is the ultimate ecological problem confronting man. Also the most difficult to solve, since the solution cannot be reconciled with the values, goals, interests, political and economic institutions prevailing in industrialized and industrializing societies."

I was talking about economic growth the other day to a young woman. And she said, "But all life is growth; that's natural, isn't it?" So I had to explain this, following Ramón Margalef and others: life moves in certain kinds of cycles, and after an occasion of disruption or turbulence, it rapidly replaces the disturbed fabric, but initially with a small number of species. As the fabric is repaired, species diversity begins to replace single-species rapid growth, and increasing complexity becomes again the model—what is called "tending toward climax," resulting in the condition sometimes called climax. That is, maximum diversity and maximum stability in a natural system. Stable because there are so many interlocking points that some particular attack on the system does not go through too many pathways, but is localized and corrected. If you have a field of nothing but grass, and grass-hoppers land on it, that's the end of your grass. If you have an acre of which grass is maybe 12 percent of the biomass, and the grasshoppers come along, you still have the other 88 percent. That's all. There's a support implicit in that, a rich-ness that is also the richness of the recycling of energy through the detritus pathways (organic matter on the downswing rather than on the upswing; the fungi, insects, and so on that live in the rotten wood and the rotten leaves rather than live off the annual production of new biomass). Detritus is a key to the stability and maturity.

Now, in Dr. Eugene Odum's terms, what we call civili-zation is an early succession phase: an immature, monocul-ture system. What we call the primitive is a mature system with deep capacities for stability and protection built into it. In fact, it seems to be able to protect itself against everything except white sugar and the money-economy trading rela-tionship; and alcohol, kerosene, nails, and matches. (It was John Stuart Mill who said, "No labor-saving invention ever really saved anybody any labor.")

One of the greatest and oldest expressions of that "prim-itive" maturity and stability is the great lore of tale and song.

So: ethnopoetics, first as a field. The politics of a new ac-ademic field. Politics of having a magazine. And the ques-tion of what we do when we start going into other people's cultures and bringing back their poems and publishing them in our magazines. I'll argue the positive side and it's simply this. An expansionist imperialist culture feels most comfortable when it is able to believe that the people it is ex-ploiting are somehow less than human. When it begins to get some kind of feedback that these people might be human beings like themselves, exploitation becomes increasingly difficult.

We can see, in a small way, how the publication—early on—of Native American poetry and narrative helped the larger white public come to realize the depth of these cultures.

Collections of American Indian mythology, folklore, and song go back to the 1880s. The quantity becomes really large after around 1900—annual reports and bulletins of the Bureau of American Ethnology, the American Ethnological Society, the memoirs and journal of the American Folklore Society, and so forth. A large body of American Indian literature in English, but almost no popular publication of it in forms that are easily available to large numbers of people. I ask why. I don't know; it may be just market economy at work, and nobody wanted to read that sort of thing. It may be that no one wanted it to be available outside a scholarly circle.

A similar case: the Ainu and the people of Japan. Dr. Kindaiichi and his associates began collecting Ainu oral literature in the 1930s, one of the largest single bodies of oral literature that's ever been collected—in Japanese translation from Ainu. I find no popular Japanese publication of any of that material through the earlier decades; in fact, it was just last year that the first easily available paperback of a selection of this oral literature came out. Until now it was buried in very expensive rare scholarly books. The attitudes of the larger Japanese public may well be transformed by this. And what will recent publication of the Villa Boas brothers' book on the Xingu do for the Brazilian Indians? A few people will read that and begin to think, "These are human beings." There is some increment of cultural and political value from the publication of oral literatures.

For most of the forty-thousand-year time span, people weren't particularly self-conscious about their own body of songs, myths, and tales, but we have some illuminating cases from the nineteenth century illustrating how publication of ethnic literature reinforced a people's sense of identity. For example, a young doctor named Elias Lönnrot set himself to walking widely through the northern parts of Finland, collecting the remaining fragments of songs and epics and tales that the people were still telling in the early nineteenth century. He strung them together in an order that he more or less perceived himself, and called it the *Kalevala*. Overnight it became the Finnish national epic and helped the Finns hold up against the Swedes on one side and the Russians on the other. It may well be that Dr. Lönnrot's walking around in the summertime is responsible for the fact that there is a nation called Finland today.

Point four in the *Alcheringa* eight-point list was "to encourage ethnologists and linguists to do work." Something happens when you do that work.

In March 1902, Alfred Kreober was in Needles, California. He wrote:

> At Ah'a-kwinyevai, in a sand-covered Mojave house, we found Inyo-Kutavére, which means 'Vanished-Pursue.' . . . He went on for six days, each of three to four hours' total narration by him and as many hours of translation by

Jack Jones and writing down by me. Each evening, he be-
lieved, I think honestly, that one more day would bring
him to the end. He freely admitted, when I asked him, that
he had never told the story through from the beginning to
the end. He had a number of times told parts of it at night
to Mojave audiences until the last of them dropped off to
sleep. When our sixth day ended he still again said another
day would see us through. But by then I was overdue at
Berkeley. And as the prospective day might once more have
stretched into several, I reluctantly broke off, promising
him and myself that I would return to Needles when I
could, not later than next winter, to conclude recording the
tale. By next winter Inyo-Kutavére had died and the tale
thus remains unfinished. . . . He was stone blind. He was
below the average of Mojave tallness, slight in figure,
spare, almost frail with age, his gray hair long and un-
kempt, his features sharp, delicate, sensitive. . . . He sat
indoors on the loose sand floor of the house for the whole of
the six days that I was with him in the frequent posture of
Mojave men, his feet beneath him or behind him to the
side, not with legs crossed. He sat still but smoked all the
Sweet Caporal cigarettes I provided. His housemates sat
around and listened or went and came as they had things to
do.[1]

That old man sitting on the sand floor telling his story is who
we must become—not A. L. Kroeber, as fine as he was.

We are all real people. Everyone on earth is a native of this
planet. All poetry is "our" poetry. Diné poetry, people poetry, Maydy poetry, human being poetry. In the forty-

thousand-year timescale we're one people. We're all equally primitive, give or take two or three thousand years here or a hundred years there. Homer then, from this standpoint, is not the beginning of a tradition but at the midpoint in a tradition. Homer incorporates and organizes the prior eight thousand years of oral material like the Chinese-trained scribes who finally put the Japanese lore in writing. Homer launches those images again forward for another couple of thousand years so that we still have Ajax cleaning powder and Hercules blasting powder. Some kind of looping.

This is the way the origin of language and poetry was described in traditional India: Brahma, the creator, is in a profound state of trance. He is silence, stillness. A thought moves somewhere in there. It manifests itself as song, the goddess Vak (the Indic muse). The goddess Vak becomes the universe itself as energy. Of that energy all subenergies are born. Vak in Indo-European philology is the same as the Latin *vox* or the English "voice." The goddess takes on another name: she's also called Sarasvati, which means "the flowing one," and she's recognized today in India as the goddess of poetry, music, and learning. She's represented as wearing a white sari, riding a peacock, carrying a vina and a scroll.

In the primal days of that energy flow, language was just "seed syllables." The practice of mantra chanting in India, the chanting of seed syllables, is a way to take yourself back to fundamental sound-energy levels. The sense of the universe as fundamentally sound and song begins poetics. They also say in Sanskrit poetics that the original poetry is the sound of running water and the wind in the trees.

In the archaic model from India (and elsewhere), there is sacred and secular song. In the case of sacred song there are again two categories: songs that are made of magic syllables and have magical meaning only, and sacred songs that have literal meanings. In the category of secular song, you can think of all the songs of all the people of the world as going through divisions like these: lullabies to sing babies to sleep; playground rhymes for kids; power vision songs of adolescent initiation; courting songs of young people; work songs—net hauling, hammer swinging, rice transplanting, canoeing, or riding; hunting songs, with a specific magical set of skills and understandings; celebration songs, war songs, and death songs. We can fit all of our own poetries into these.

One other category that is most important could be called "healing songs." Those who received particularly strong power vision songs became the medicine persons, the singer-healers. They come to us in history as the fellows Plato wanted to kick out. I think that a concern for nature and the integrity of the many realms of critters is a long and deeply rooted concern of the poet. The job of the singer was to sing the voice of corn, the voice of the Pleiades, the voice of bison, the voice of antelope. To contact in a very special way an "other" that was not within the human sphere; something that could not be learned by continually consulting other human teachers, and could only be learned by venturing outside the human borders and going into your own mind wilderness, unconscious wilderness. Thus, poets were always "pagans," which is why Blake said Milton was of the devil's party but he didn't know it. The devil is, after all, not the devil at all; he is the miming elk shaman dancer at Trois Frères, with antlers and a pelt on his back, and what he's doing has to do (we may speculate) with animal fertility in the springtime, with human-animal communication. Of the archaic array of song genres, we can say that "healing songs" were the most difficult, and the most profound. They are approached in the old question, How do you prepare your mind to become a singer? It takes an attitude of openness, inwardness, gratitude; plus meditation, fasting, a little suffering, some rupturing of the day-to-day ties with the social fabric. I quote from the Papago (or more properly, O'odham):

> *A man who desires song did not put his mind on words and tunes. He put it on pleasing the supernaturals. He must be a good hunter or a good warrior. Perhaps they would like his ways. And one day in natural sleep he would hear singing. He hears a song and he knows it is the hawk singing to him of the great white birds that fly in from the ocean. Perhaps the clouds sing or the wind or the feathery red rain spider on its invisible rope. The reward of heroism is not personal glory nor riches. The reward is dreams. One who performs acts of heroism puts himself in contact with the supernatural. After that, and not before, he fasts and waits for a vision. The Papago holds to the belief that visions do not come to the unworthy, but to the worthy man who shows himself humble there comes a dream and the dream always contains a song.*[2]

To go back to the lore of the muse: the image of the muse, the goddess, is strong in our occidental tradition, and it's also present in the Sanskrit and Tamil traditions of India. The Chinese tradition very early had a muse point of view that became covered over. It's to be found in Taoism, and within the emphasis on the female, the feminine, the spirit of the valley, the yin—Taoism being, following Dr. Joseph Needham's assessment of it in *Science and Civilization in China*, the largest single coherent chunk of matrilineal-descent, mother-consciousness-oriented, Neolithic culture that went through the, so to speak, sound barrier of civilization into the Iron Age and came out the other side halfway intact. Thus, through its whole political history Taoism has been antifeudal and antipatriarchal. Taoism has an earth goddess half-visible in its earliest manifestations. All these old poetries give us a poetics of the earth.

Concentrations of communication energy result in language; certain kinds of compressions of language result in mythologies; compression of mythologies brings us to songs. "The transmission"—this is Dr. H. T. Odum, brother of Dr. Eugene Odum—"of information is an important part of any complex system. Small energy flows that have high amplification factors have value in proportion to the energies they control. As the smallest of energy flows, information pathways may have the highest value of all when they open work-gate valves on power circuits. The quality of this information, tiny energies in the right form, is so high that in the right control circuit it may obtain huge amplifications and control vast power flows."[3] In the great universe, the main "theme" of energy flow is in massive objects coming together and realizing their own gravity. Solar radiation per square meter out in space is 1.395. Solar energy comprises

99.98 percent of the energy influx on the earth. The tiniest fraction of that is captured by the chlorophyll of plant leaves. Here's the poetics, as articulated by biologist Lewis Thomas:

> *Morowitz has presented the case, in thermodynamics, for the hypothesis that a steady flow of energy from the inexhaustible source of the sun to the unfillable sink of outer space, by way of the earth, is mathematically destined to cause the organization of matter into an increasingly ordered state. The resulting balancing act involves the ceaseless clustering of bonded atoms into molecules of higher and higher complexity and the emergence of cycles for the storage and release of energy. In a nonequilibrium steady state, which is postulated, the solar energy would not just flow to the earth and radiate away; it's thermodynamically inevitable that it must rearrange matter into symmetry, away from probability, against entropy, lifting it so to speak into a constantly changing condition of rearrangement and molecular ornamentation. If there were to be sounds to represent this process, they would have the arrangement of the Brandenburg concertos, but I'm open to wondering whether the same events are recalled by the rhythms of insects, the long pulsing runs of bird songs, the descants of whales, the modulated vibrations of millions of locusts in migration.*[4]

That is, on some subliminal level, what we're tuned into— for our language, for our songs.

The rhythms and modulations of archaic music, the little chants and choruses of songs and dances, the trancelike repetitions and dawn-clear images, the borrowed calls of birds, and the metaphoric rush of streams that are in the weave of world poetics all manifest what Lewis Thomas found in the Brandenburg concertos. The study of ethnopoetics provides would-be poets and scholars with hundreds of models of poem strategies, forms, angles of imagery, types of wordplay and mindplay that enrich our sense of human accomplishment without stealing anything from anybody.

The way to express gratitude and respect for these teachings of poetry, music, and song is to join in the work of helping your nearest endangered subsistence society in its struggle against the rape of land and culture. In terms of life in the late twentieth century, with all its terrible suffering, the indigenous subsistence people, their cultures, and their home jungles or forests, have fared least well. They are disappearing, even as we praise their songs.

Notes

1. A. L. Kroeber, "A Mojave Historical Epic," Anthropological Records, *vol. 11, no. 2 (Berkeley: University of California Press, 1951),* *71.*

2. *Ruth Underhill,* Singing for Power *(Berkeley: University of California Press, 1968), 7.*

3. *H. T. Odum,* Environment, Power, and Society *(New York: Wiley, 1971), 72.*

4. *Lewis Thomas,* The Lives of a Cell *(New York: Viking Press, 1974), 27–28.*

[Some of this was originally given as a talk at the ethnopoetics conference held at the University of Wisconsin at Milwaukee in April 1975. Michael Benamou, Jerry Rothenberg, and Dennis Tedlock were the organizers of this event. An earlier version was published in The Old Ways *(San Francisco: City Lights, 1977).]*

The Incredible Survival
of Coyote

I

Of all the echoes of Native American lore in modern poetry, the continuing presence of Coyote is the most striking. Poets are looking back now at the history of the West. They are making reference to almost as much Native American lore as they are drawing on, say, the folklore of the cowboy or the mountain men. Why is this? When the early fur traders began to explore the Great Basin country in the 1820s they made contact with some very hardy and clever nations that had ancient techniques for living in those arid regions. Only a few bison strayed into eastern Utah or Idaho, and horses were just beginning to catch on. But these nations from the Rockies westward—Great Basin, the northern Interior Plateau, the Southwest, and on to California—have given the world the rich lore of Old Man Coyote. And now this potent character has made himself totally at home in contemporary poetry and art.

Coyote the critter survives where the wolf is almost extinct throughout the West because it took no poison bait. Strychnine-laced cow carcasses that ranchers put out for the wolf did the wolf in. The coyotes learned very early not to take poisoned bait, and they still flourish. Similarly, from the Rockies westward and from Mexico north well into Canada, you find the trickster hero Coyote to be one of the most prominent elements of native culture.

There are lots of stories about Coyote, or Coyote Man as he's called to distinguish him from coyote the animal. Old Man Coyote lived in myth time, the dreamtime—and lots of things happened then. Over on the other side of the Cascades, the trickster is called Raven. In the Great Lakes region, sometimes he's called Hare, but out here it's Coyote Man.

He's always traveling, he's really stupid, he's kind of bad—in fact, he's really awful, he's outrageous. But he's done some good things, too: he got fire for people. The Mescaleros say he found where the fire was kept. It was kept by a bunch of flies in a circle, and he couldn't get into the circle, but he was able to stick his tail in there and get his tail burning, and then off he scampered and managed to start some forest fires with his tail, and the fire he started kept running around the world, and people are still picking it up here and there. He taught people up by the Columbia River how to catch salmon. He taught people which plants were edible. So he's done some good things.

But most of the time he's just into mischief. It's Coyote's

fault that there's death in the world—this from the Maidu in California. Earthmaker made the world so that people wouldn't get old, wouldn't die. He made a lake so that if people began to feel they were getting old, they could get into this lake and grow young again. He also made it so that every morning when you woke up, you could reach outside your lodge door and there'd be a bowl of hot, steaming acorn mush to eat. People didn't have to work for food in those days. But Coyote went around agitating the human beings, saying, "Now, you folks, don't you think this is kind of a dull life? There ought to be something happening here; maybe you ought to die." And they'd say, "What's that, death?" And he'd say, "Well, you know, if you die, then you really have to take life seriously, you have to think about things more." He kept agitating this way, and Earthmaker heard him. Earthmaker shook his head and said, "Oh, boy, things are going to go all wrong now." And Coyote kept talking this death idea around, and pretty soon things started happening. The people were having a footrace. Coyote Man's son got out there, and by golly, he stepped on a rattlesnake and the rattlesnake bit him, and he fell over and lay on the ground, and everybody thought he was asleep for the longest time. Coyote kept shouting, "Wake up, come on now, run." Finally Earthmaker looked at him and said, "You know what happened? He's dead. You asked for it." And Coyote said, "Well, I changed my mind, I don't want people to die after all, now let's have him come back to life." But Earthmaker said, "It's too late now, it's too late now."

There are plenty of stories about Coyote Man and his

sheer foolishness. For instance, he's out walking along, and he sees these beautiful little gold-colored cottonwood leaves floating down to the ground, and they go this . . . this . . . this . . . this . . . this, this, this, this, and he just watches them for the longest time. Then he goes up and he asks those leaves, "Now, how do you do that? That's so pretty the way you come down." And they say, "Well, there's nothing to it, you just get up in a tree, and then you fall off." So he climbs up the cottonwood tree and launches himself off, but he doesn't go all pretty like that, he just goes crash and kills himself. But Coyote never dies; he gets killed plenty of times, he comes back to life again, and he goes right on traveling.

Another time, Coyote went to the world above this one, and the only way to get back was to come down a spiderweb. And the spider told him, "Now, when you go down that spiderweb, don't look down, and don't look back; just keep your eyes closed until your feet hit the bottom, and then you'll be OK." So Coyote's going down this spiderweb, and he's getting kind of restless, and he says, "Well, now, I'm just sure I'm about to touch bottom now, any minute my foot's going to touch bottom, I'm going to open my eyes." And he opens his eyes, and naturally the spiderweb breaks, and he falls and kills himself. So he lies there, and the carrion beetles come and eat him, and some of his hair blows away, and pretty soon his ribs are coming out. About six months go by and he is really looking messy, but he begins to wake up. And he opens one eye, and he tries to open the other, but he can't find the other eye, so he reaches out and sticks a pebble in his eye socket, and then a blue jay comes along and puts a little pine pitch on the pebble, and then Coyote can see through that. And he pulls himself back together and goes to look for a couple of his ribs that have kind of drifted down the hill, and he puts them in place and says, "Well, now I'm going to keep on traveling."

He gets into some awful mischief. There's a story we find all the way from the Okanogan country right down into the Apache country about the time Coyote Man got interested in his oldest daughter. He wanted to have an affair with his oldest daughter, and he couldn't get over the idea, so finally one day he tells his family, "Well, I'm going to die now," or in some versions he says, "I'm going on a long trip." And he just goes to the other side of the hill a ways, and he puts on a disguise. But before he goes, he tells everyone, "If a man comes along, some handsome man with a big satchel who has a couple of jackrabbits to offer you, now you be real nice to him, let him in, take good care of him." So then he goes off around the hill, puts on his disguise, picks up a couple of jackrabbits, and comes back, and the girls say to their mother, "Why, sure enough, here's a man just like Daddy said we should be nice to." And so they take him into the house and feed him, and he sits in the corner and talks in kind of a muffled voice, and then he starts saying, "That sure is a pretty oldest daughter you've got there. Wouldn't you like to marry her to somebody?" And he's almost going to get away with this, but then he asks them to go over his hair for lice and so they're checking him and they find this scar on one cheek, and say, "Why, it's just Daddy." Stories like that.

II

How did this lore find its way into recent literature? I was hitchhiking in 1951 from San Francisco to Indiana University to enter graduate school in anthropology. I got picked up by some Paiute men just outside of Reno in the evening. We drove all night to Elko. It turned out they had a couple cases of beer in the car, so we were doing that old Far Western recreation of drinking and driving a couple hundred miles. They had had jobs as steelworkers in Oakland and now were heading back home. They started telling Coyote tales. (They also talked about Jesus. One said Jesus was a great gambler; he was perhaps the best gambler in the United States.) They told these Coyote stories with obvious relish, as part of their world, but also with that distance that comes from having been urban, from having worked their steelworking jobs and having been to BIA (Bureau of Indian Affairs) school. But still they were telling them.

Another way it happened is this. In my own poetry, back twenty years or so, Coyote began to crop up here and there. I think I was one of the first to invoke him, but I wasn't the only one. There was a literary magazine called *Coyote's Journal*, edited by James Koller, in the sixties and early seventies. Even earlier, Jaime de Angulo wrote down the stories he had told to his children, and they were published in the middle fifties as *Indian Tales*. Jaime de Angulo was originally a Spanish medical doctor who converted to being an anthropologist and linguist in the Southwest and then in California during the twenties and thirties. He became an anarchist-

bohemian culture hero in San Francisco and Big Sur in post–
World War II days. He was a friend of Robinson Jeffers; in
fact, I have heard it said that he was the only human being
that Robinson Jeffers would let into his house at any time of
the day or night. He appears in some of Jeffers's poems as the
"Spanish cowboy," because at one time de Angulo was running a working ranch on Partington Ridge up above Big Sur.
Anyway, he had direct contact with Coyote and other American Indian lore as a linguist and anthropologist, and his *Indian Tales* had a direct influence on writers in the Bay Area.
Some of those writers produced a small body of poetry that
refers to Coyote as though he were familiar to everyone.

To understand why Coyote became so familiar we should
look at the West again. We keep asking, "What is a westerner, really, and what is the West?" The answer keeps shifting. I'll offer my interpretation, hoping that I have some
credentials because I come from an old western family in the
logging country. For years the literature of the West was
concerned with exploitation and expansion. This is what
we mean when we talk about the "epic" or "heroic" period
—a time of rapid expansion, of first-phase exploitation.
This literature is not a literature of place. It is a history
and a literature of feats of strength and of white, English-
speaking-American human events. It's only about this place
by accident. The place figures in it as an inhospitable and unfamiliar terrain; Anglos from temperate climates are suddenly confronted with vast, treeless, arid spaces. Space and
aridity, confronting them and living with them, are themes
in western literature, but only incidentally. It could just as
well be an Icelandic saga, or a heroic epic of Indo-European

people spreading with their cattle and wagons into any new and unfamiliar territory, as they did in 1500 B.C. when they moved down into the basin of the Ganges or into Anatolia. The white settlement of the West presented us with images of manliness, vigor, courage, humor, and heroics. These became a strong part of our national self-image—perhaps the strongest part, the most pervasive, the one that has been exported to the rest of the world. There are southern images (Daniel Boone); there are the Yankee and several other archetypes in American folk literature. But the western image, a kind of amalgam of mountain man, cowboy, and rancher, is probably the strongest. The West ceases to exist (whether an area is geographically western or not) when the economy shifts from direct, rapid exploitation to a stabilized agricultural base. Heroics go with first-phase exploitation: the fur trade, then the cattle industry, then mining, then logging.

I grew up in western Washington, which some people would perhaps exclude from the "western literature" sphere since it is not arid and treeless. But it lies within the myth of the West because it took part in the direct-exploitation economy, and it has all the tall tales, the energy, the unpredictability, the mobility, and the uprootedness that go with that kind of work. Now, oil fields are similarly "western," since they still have that rough-and-tumble angle to them, and that means Alaska is also the West.

Another aspect of the West was that men were removed some distance from women. The literary critic Leslie Fiedler has written on this, saying that one aspect of the heroic and epic West is that men got away from home and away from women. The men were also removed from their fathers, and they were beyond the reach of the law—the patriarchal figure of the nation-state. Thus, the West was being occupied by boys without fathers and mothers, who were really free to get away with things for a while. And that's why there is so much raw humor in the lore of the West.

But something has happened to our sense of the West since World War II. I can see a bit of how it happened in me. Western lore has been changing from a story of exploitation and expansion by white people into a quest for a sense of place. Those early westerners did not know where they were—except for the mountain men, who became almost Indians.

The early settlers were on their way to a sense of place. I know that my grandmother was able to cook with a few wild plants in Kitsap County, Washington, and get wild blackberries and a few edible mushrooms from the forest. But my grandmother's generation was the last; the next generation grew up with supermarkets and canned food. The potential for a viable self-sufficient rural West evaporated after about one generation. What we have now in the West is a kind of semiurban population with everyone driving hundreds of miles for food and work.

So current writers and many young people look to Native American lore. There is something to be learned from the Native American people about where we all are. It can't be learned from anybody else. We have a western white history of 150 years, but the Native American history (the datings are always being pushed back) was first ten thousand years, then it was sixteen thousand years, then people started talking about thirty-five thousand years. Now they've found charred mammoth bones in Santa Barbara that look like they're from a Paleolithic barbecue. When we read a tale or look at a little bit of American Indian folklore or myth, we're seeing just the tip of the iceberg of forty or fifty thousand years of human experience on this continent, in this place. It takes a great effort of imagination to enter into that, to draw from it, but there is something powerful there.

III

Much carefully gathered Native American myth and lore resides in our libraries in the form of the bulletins and reports of the Bureau of American Ethnology, going back to the 1880s. Respectable, scrupulous, careful collections, they are for the most part unbowdlerized. (In the early days they took the more scatological sections and translated them into Latin, but then everybody knew Latin in those days so. . . . In recent years they haven't done that.) That is what I learned from, and there is an irony in that, too. I grew up knowing Puget Sound Indians, but the way I first learned their literature was by going to the library. Franz Boas, Edward Sapir, John Swanton, Melville Jacobs, Thelma Jacobson, Alfred Lewis Kroeber and his students, Harry Hoijer, M. E. Opler—almost all of them in one way or another disciples of Franz Boas—gave decades of their lives to the collecting of texts in the original languages, with translations into English, of the lore of every cultural group that they could make contact with from British Columbia south. It is an extremely rich body of scholarship, and some of us learned how to draw it into our own work, to enjoy it—and learn from it a new way to look at nature. One of the first things that excited me about Coyote tales was the raw Dadaistic energy, leaping somehow into a modern frame of reference. There's nothing as useful as the direct transcription, as literally close as possible to the original text in whatever language it was, Kwakiutl or Mescalero Apache. The true flavor will still be there. (There is a perennial argument about whether you get more out of something when somebody has made it more readable, more literary, taken the brackets and the parentheses out, and the dots and the ellipses and the footnotes; cleaned it up. Is that better, or is it better when you get down to the primary source and try to use that? Well, I'm all for the primary source in historical materials, whether American Indian or otherwise, because I would prefer to do the editing with my own imagination rather than let somebody else's imagination do the work for me. At least then, if there are errors in interpretation, they're my own errors and not somebody else's.)

Another way to draw nigh to Coyote is the direct experience of being in the big country. I spent a lot of time in eastern Oregon. Sure, stories about the early stages and wagon trains going across eastern Oregon, tales of the early wheat ranchers—they're interesting, but they don't help with the sense of place. Coyote stories in the Wasco and Wishram texts began to teach me something about the real flavor of the land, began to move me back just a trifle from historical time into myth time, into geological time. Looking forward, then, I can only speculate that (to finish with comments on the West) future North American people are bound to become increasingly concerned with *place*. This will not be the same as regionalism. Regionalism in the past has been a human history: the story of particular human habits and oddities and quirks—ethnic diversities and whatever—that are established in a region. That's the human history of newcomers, and it's often very good literature, but it's not tuned to the spirit of place as I think it will be in the future. Ecologists and economists tell us that harmony with the local place and a way of life that does not exhaust the resources and that can be passed on to your children and grandchildren with no fear of depletion are how we must learn to live. As people come to understand this, they will look back with few regrets and say that the heroic period of the West was entertaining, but we've been learning a lot from the Native Americans since then. And for the Indian peoples western history is not a glorious epic history with nice tall tales on the side; it's a history of humiliation, defeat, and dispossession.

Coyote was interesting to me and my colleagues because he spoke to us of place, and became almost like a guardian, a protector spirit. The other feeling—our fascination with the trickster—has to come out of something inside us. For me I think the most interesting psychological aspect of the trickster stories is that they mix up good and evil. Coyote clearly manifests benevolence, compassion, helpfulness to human beings, and has a certain dignity; and on other occasions he is an utter fool. But old Coyote Man, he's just always traveling along, doing the best (worst?) he can.

Growing up in the fifties in Portland, Oregon, going to Reed College and associating with still-struggling radical professors who had found somewhere to teach, I drew on the

IWW lore of my grandfather, the grass-roots political radicalism of the Northwest. The West was heroics, but in the fifties and sixties we didn't feel like heroics, and because the trickster presents himself to us as an antihero, he was immediately an attractive figure. For the same reasons, you find antiheroics in the writings of post–World War II French and Italian or English writers. Artaud is a trickster; William Burroughs in his novels talks out of the side of his mouth with a kind of half-Coyote, half–Dashiell Hammett dry style. The trickster image is basic; it has to do with turning heroes into a joke and with celebrating nimble wit, craftiness, and foresight. There's a white American frontier storytelling that is somewhat like Coyote lore—irresponsible, humorous, and unpredictable. Mike Fink tall tales were being made up while men scrapped and boasted with each other, the Indians telling their Coyote tales and the white men telling their tall tales. (Of course, I'm only reading Coyote as I can, namely as a twentieth-century, West Coast white American. How the Native American people themselves actually saw Coyote is another question.)

When the Coyote figure comes into modern American poetry, it is not just for a sense of place. It is also an evocation of the worldwide myth, tale, and motif storehouse. Poetry has always done this—drawing out, re-creating, subtly altering for each time and place the fundamental images.

The crazy yaps and howls of Coyote the critter are also a gate to Coyote the trickster. Coyote the animal is a perfect expression of a specific set of natural relationships, as appropriate to the mountains and deserts of the West as the trickster image is appropriate to certain human needs. It's a marvelous coming together—the meeting of trickster and coyote. Coyote is smart, quick, omnivorous, careful, playful; a good parent; opportunistic and graceful. Contemporary people who know nature by direct observation have seen all this; the old-time Native Americans knew even more. There are specific things to be learned from each bird, plant, and animal—a natural system is a total education—and this learning is moral, as well as being useful for survival. The red-tailed hawk teaches us to have a broad view of things, while not missing the stirring of a single mouse.

IV

The Trickster is probably the most archaic and widely diffused figure in world folklore. No wonder his name is often "Old Man." He is the Old One, the Ancient Buddha. We can wonder what the Trickster meant to our Upper Paleolithic ancestors; with a laugh and a shiver we almost intuitively know—and can guess at the worldview (human and nonhuman lives entwined) projected from the most archaic constellation of "Lady of the Mammoth" heavy-bodied Earth Mother, Trickster Old Man Jackal/Fox/Coyote, the Great Bear of the Mountains, and the Dancer with Antlers.

In their quiet, conservative corner of the globe, the Great Basin Indians and the nations of California were, a century ago, still living and transmitting an international body of lore—the same lore that is the foundation of the "high" literatures of India, the Middle East, the Mediterranean, and western Europe. And just as planetary humankind becomes "ecosystem"-oriented again, so the most sophisticated and agonized contemporary theologies come close to confessing that God must be a trickster.

But God is not exactly a trickster. Coyote the animal, human being the animal, bear the animal are (the Ainu would say) just *hayakpe*—"armor" or masks, or food to be served, shapes and functions assumed in the service of Great Nature. We slip those masks a bit to the side and see there Coyote Man the trickster; Bear, the king of the mountains; Deer Mother, queen of compassion. In turn, those Type Beings, mind-created, earth-created, are also illusions. The Shining One peeks out from behind a boulder and is gone—is always there. But all this archetypal stuff is not really our concern. Our concern is the kids sleeping in the back room, snow on the far hills, a coyote howling in the sagebrush moonlight.

[This essay is based on a talk given at the 1974 Western Writers Conference, Utah State University, Logan. An earlier version was published in The Old Ways *(San Francisco: City Lights, 1977).]*

Unnatural Writing

"*N*ature writing" has become a matter of increased literary interest in the last few years. The subject matter "nature," and the concern for it (and us humans in it), have come—it is gratifying to note—to engage artists and writers. This interest may be another strand of postmodernism, since the modernist avant-garde was strikingly urban-centered. Many would-be writers approach this territory in a mode of curiosity, respect, and concern, without necessarily seeking personal gain or literary reputation. They are doing it for love—and the eco-warrior's passion, not money. (There is still a wide range of views and notions about what nature writing ought to be. There is an older sort of nature writing that might be seen as largely essays and writing from a human perspective, middle-class, middlebrow Euro-American. It has a rhetoric of beauty, harmony, and sublimity. What makes us uncomfortable sometimes with John Muir's writing is an excess of this. He had contemporaries, now forgotten, who were far worse.)

Natural history writing is another branch. Semiscientific, objective, in the descriptive mode. Both these sorts are "naively realistic" in that they unquestioningly accept the front-mounted bifocal human eye, the poor human sense of smell, and other characteristics of our species, plus the assumption that the mind can, without much self-examination, directly and objectively "know" whatever it looks at. There has also always been a literature of heroic journals and adventure. And there is an old mix of science, nature appreciation, and conservation politics that has been a potent part of the evolution of the conservation movement in the United States. The best of this would be seen in the work of Rachel Carson and Aldo Leopold. All of these writings might be seen by some as mildly anthropocentric, but the work is worthy and good-hearted. We are in its debt.

Nature writing has been a class of literature held in less than full regard by the literary establishment, because it is focused on something other than the major subject matter of mainstream occidental writing, the moral quandaries, heroics, affairs of the heart, and soul searchings of highly gifted and often powerful people, usually male. Tales of the elites. In fact, up until a decade ago nature writing was relegated pretty much to a status like that of nineteenth-century women's writing—it was seen as a writing of sensibility and empathy and observation, but off to the side, not really serious, not important.

But if we look at the larger context of occidental history, educated elites, and literary culture, we see that the natural world is profoundly present in and an inescapable part of the great works of art. The human experience over the larger part of its history has been played out in intimate relationship to the natural world. This is too obvious even to say, yet it is often oddly forgotten. History, philosophy, and literature naturally foreground human affairs, social dynamics, dilemmas of faith, intellectual constructs. But a critical subtheme that runs through it all has to do with defining the human relationship to the rest of nature. In literature, nature not only provides the background, the scene, but also many of the characters. The "classical" world of myth is a world in which animal beings, supernatural figures, and humans are actors and interacters. Bears, bulls, and swans were not abstractions to the people of earlier times but real creatures in very real landscapes. The aurochs—the giant wild cow, *Bos primigenius*, who became Zeus to Europa—survived in pockets of the European forests until medieval times.

In *The Practice of the Wild* (San Francisco: North Point Press, 1990), I point out that through most of human history

populations were relatively small and travel took place on foot, by horse, or by sail. Whether Greece, Germania, or Han China, there were always nearby areas of forest, and wild animals, migratory waterfowl, seas full of fish and whales, and these were part of the experience of every active person. Animals as characters in literature and as universal presences in the imagination and in the archetypes of religion are there because they were there. *Ideas and images of wastelands, tempests, wildernesses, and mountains are born not of abstraction but of experience: cisalpine, hyperboreal, circumpolar, transpacific, or beyond the pale. [This is the world people lived in up until the late nine-*

teenth century. Plentiful wildlife, open space, small hu-
man population, trails instead of roads—and human lives
of individual responsibility and existential intensity. It is
not "frontier" that we're considering, but the Holocene
era, our present era, in all its glory of salmon, bear, elk,
deer, and moose.] Where do the sacred salmon of the Celts,
the Bjorns and Brauns and Brun-(hilde)-s [bher = bear]
of northern European literature, the dolphins of the Med-
iterranean, the Bear dances of Artemis, the Lion skin of
Herakles come from but the wild systems the humans lived
near?

> *Those images that yet*
> *Fresh images beget*
> *That dolphin-torn, that gong-tormented sea.*

Many figures in the literary field, the critical establishment, and the academy are not enthralled with the natural
world, and indeed some positively doubt its worth when
compared to human achievement. Take this quote from
Howard Nemerov, a good poet and a decent man:

> *Civilization, mirrored in language, is the garden where*
> *relations grow; outside the garden is the wild abyss.*

The unexamined assumptions here are fascinating. They
are, at worst, crystallizations of the erroneous views that enable the developed world to displace Third and Fourth
World peoples and overexploit nature globally. Nemerov
here proposes that language is somehow implicitly civilized
or civilizing, that civilization is orderly, that intrahuman relations are the pinnacle of experience (as though all of us,

and all life on the planet, were not interrelated), and that "wild" means "abyssal," disorderly, and chaotic.

First take language. Some theorists have latched onto "language" as that which somehow makes us different. They have the same enthusiasm for the "Logos" as the old Summer Institute of Linguistics had for Bible translation into unwritten languages. In fact, every recent writer who doesn't know what else to say about his or her work—when asked to give a sound bite—has declared, "Well, I'm just fascinated with language." The truth is language is part and parcel of consciousness, and we know virtually nothing about either one. Our study and respect should extend to them both.

On another tack, the European deconstructionists assume, because of their monotheistic background, that the Logos died along with God. Those who wish to decenter occidental metaphysics have begun to try to devalue both language and nature and declare them to be further tools of ruling-class mythology. In the past, the idea that the external world was our own invention came out of some variety of idealist thought. But *this* version leads to a totally weird philosophical position that, since the proponents are academic "meta-Marxists," might be called "materialist solipsism." But they are just talk.

There is some truly dangerous language in a term heard in some business and government circles: "sustainable development." Development is not compatible with sustainability and biodiversity. We must drop talking about development and concentrate on how to achieve a steady state condition of real sustainability. Much of what passes for economic development is simply the further extension of the destabilizing, entropic, and disorderly functions of industrial civilization.

So I will argue that consciousness, mind, imagination, *and* language are fundamentally wild. "Wild" as in wild ecosystems—richly interconnected, interdependent, and incredibly complex. Diverse, ancient, and full of information. At root the real question is how we understand the concepts of order, freedom, and chaos. Is art an imposition of order on chaotic nature, or is art (also read "language") a matter of discovering the grain of things, of uncovering the measured chaos that structures the natural world? Observation, reflection, and practice show artistic process to be the latter.

Our school-in-the-mountains here at Squaw Valley is called "art of the wild." (I was wondering just what edible root might have been growing so profusely in this wet mountain bottomland to have caused it to be called "Squaw Valley." Any place with the word *squaw* in the name is usually where some early trappers saw numerous Native American women at work gathering wild food; here it might have been *Brodiaea* bulbs. This naming practice is as though some native women coming on a Euro-American farming community had called it White Boy Flats.)

The "art of the wild" is to see art in the context of the process of nature—nature *as* process rather than as product or commodity—because "wild" is a name for the way that phenomena continually actualize themselves. Seeing this also serves to acknowledge the autonomy and integrity of the nonhuman part of the world, an "Other" that we are barely beginning to be able to know. In disclosing, discovering, the wild world with our kind of writing, we may find ourselves breaking into unfamiliar territories that do not seem anything like what was called "nature writing" in the past. The work of the art of the wild can well be irreverent, inharmonious, ugly, frazzled, unpredictable, simple, and clear—or virtually inaccessible. Who will write of the odd barbed, hooked, bent, splayed, and crooked penises of nonhuman male creatures? Of sexism among spiders? Someone will yet come to write with the eye of an insect, write from the undersea world, and in other ways that step outside the human.

In *Practice* it says:

Life in the wild is not just eating berries in the sunlight. I like to imagine a "depth ecology" that would go to the dark side of nature—the ball of crunched bones in a scat, the feathers in the snow, the tales of insatiable appetite. Wild systems are in one elevated sense above criticism, but they can also be seen as irrational, moldy, cruel, parasitic. Jim Dodge told me how he had watched—with fascinated horror—orcas methodically batter a gray whale to death in the Chukchi Sea. Life is not just diurnal and a property of large interesting vertebrates, it is also nocturnal, anaerobic, cannibalistic, microscopic, digestive, fermentative: cooking away in the warm dark. Life is well maintained at a four mile ocean depth, is waiting and sustained on a frozen rock wall, and clinging and nourished in hundred-degree desert temperatures. And there is a world of nature on the decay side, a world of beings who do rot and decay in the shade. Human beings have made much of purity, and are repelled by blood, pollution, putrefaction. The

other side of the "sacred" is the sight of your beloved in the
underworld, dripping with maggots. Coyote, Orpheus,
and Izanagi cannot help but look, and they lose her.
Shame, grief, embarrassment, and fear are the anaerobic
fuels of the dark imagination. The less familiar energies of
the wild world, and their analogs in the imagination, have
given us ecologies of the imagination. . . .

Narratives are one sort of trace that we leave in the
world. All our literatures are leavings, of the same order as
the myths of wilderness peoples who leave behind only
stories and a few stone tools. Other orders of beings have
their own literatures. Narrative in the deer world is a track
of scents that is passed on from deer to deer, with an art of
interpretation which is instinctive. A literature of blood-
stains, a bit of piss, a whiff of estrus, a hit of rut, a scrape
on a sapling, and long gone. And there might be a "narra-
tive theory" among these other beings—they might rumi-
nate on "intersexuality," or "decomposition criticism."

I propose this to turn us loose to think about "wild writing" without preconception or inhibition, but at the same
time with craft. The craft could be seen as the swoop of a
hawk, the intricate galleries of burrowing and tunneling under the bark done by western pine bark bettles, the lurking
at the bottom by a big old trout—or the kamikaze sting of a
yellow jacket, the insouciant waddle of a porcupine, the constant steadiness of a flow of water over a boulder, the chatter
of a squirrel, hyenas moaning and excavating the bowels of
a dead giraffe under a serene moon. Images of our art. Nature's writing has the potential of becoming the most vital,

radical, fluid, transgressive, pansexual, subductive, and morally challenging kind of writing on the scene. In becoming so, it may serve to help halt one of the most terrible things of our time—the destruction of species and their habitats, the elimination of some living beings forever.

Finally, let us not get drawn too far into dichotomous views and arguments about civilization versus nature, the domesticated versus the wild, the garden versus the wild abyss. Creativity draws on wildness, and wildness confers freedom, which is (at bottom) the ability to live in the real physical daily world at each moment, totally and completely.

Some Points for a "New Nature Poetics"

- That it be literate—that is, nature literate. Know who's who and what's what in the ecosystem, even if this aspect is barely visible in the writing.
- That it be grounded in a place—thus, place literate: informed about local specifics on both ecological-biotic and sociopolitical levels. And informed about history (social history and environmental history), even if this is not obvious in the poem.
- That it use Coyote as a totem—the Trickster, always open, shape shifting, providing the eye of other beings going in and out of death, laughing with the dark side.
- That it use Bear as a totem—omnivorous, fearless, without anxiety, steady, generous, contemplative, and relentlessly protective of the wild.

- That it find further totems—this is the world of nature, myth, archetype, and ecosystem that we must each investigate. "Depth ecology."
- That it fear not science. Go *beyond* nature literacy into the emergent new territories in science: landscape ecology, conservation biology, charming chaos, complicated systems theory.
- That it go further with science—into awareness of the problematic and contingent aspects of so-called objectivity.
- That it study mind and language—language as wild system, mind as wild habitat, world as a "making" (poem), poem as a creature of the wild mind.
- That it be crafty and get the work *done*.

[The original version of this essay was given as a talk the first year of the "Art of the Wild" nature-writing conference series, held at Squaw ("Brodiaea Harvesters") Valley in July 1992.]

Language Goes Two Ways

*L*anguage has been popularly described, in the Occident, as that by which humans bring order to the "chaos of the world." In this view, human intelligence flowers through the supposedly unique faculty of language, and with it imposes a net of categories on an untidy universe. The more objective and rational the language, it is thought, the more accurate this exercise in giving order to the world will be. Language is considered by some to be a flawed mathematics, and the idea that mathematics might even supplant language has been flirted with. This idea still colors the commonplace thinking of many engineer types and possibly some mathematicians and scientists.

But the world—ordered according to its own inscrutable mode (indeed a sort of chaos)—is so complex and vast on both macro and micro scales that it remains forever unpredictable. The weather, for hoary example. And take the very mind that ponders these thoughts: in spite of years of personhood, we remain unpredictable even to our own selves.

Often we wouldn't be able to guess what our next thought will be. But that clearly does not mean we are living in hopeless confusion; it only means that we live in a realm in which many patterns remain mysterious or inaccessible to us.

Yet we can affirm that the natural world (which includes human languages) is mannerly, shapely, coherent, and patterned *according to its own devices.* Each of the four thousand or so languages of the world models reality in its own way, with patterns and syntaxes that were not devised by anyone. Languages were not the intellectual inventions of archaic schoolteachers, but are naturally evolved wild systems whose complexity eludes the descriptive attempts of the rational mind.

"Wild" alludes to a process of self-organization that generates systems and organisms, all of which are within the constraints of—and constitute components of—larger systems that again are wild, such as major ecosystems or the water cycle in the biosphere. Wildness can be said to be the essential nature of nature. As reflected in consciousness, it can be seen as a kind of open awareness—full of imagination but also the source of alert survival intelligence. The workings of the human mind at its very richest reflect this self-organizing wildness. So language does not impose order on a chaotic universe, but reflects its own wildness back.

In doing so it goes two ways: it enables us to have a small window onto an independently existing world, but it also shapes—via its very structures and vocabularies—how we see that world. It may be argued that what language does to our seeing of reality is restrictive, narrowing, limiting, and possibly misleading. "The menu is not the meal." But rather than dismiss language from a spiritual position, speaking vaguely of Unsayable Truths, we must instead turn right back *to* language. The way to see *with* language, to be free with it and to find it a vehicle of self-transcending insight, is to know both mind and language extremely well and to play with their many possibilities without any special attachment. In doing this, a language yields up surprises and angles that amaze us and that can lead back to unmediated direct experience.

Natural Language, with its self-generated grammars and vocabularies constructed through the confusion of social history, expresses itself in the vernacular. Daily usage has many striking, clear, specific usages and figures of speech that come through (traditionally) in riddles, proverbs, stories, and such—and nowadays in jokes, raps, wildly fluid slang, and constant experiment with playful expressions (the dozens, the snaps). Children on the playground chant rhymes and enjoy fooling with language. Maybe some people are born with a talent for language, just as some people are born with a talent for math or music. And some natural geniuses go beyond being street singers, mythographers, and raconteurs to become the fully engaged poets and writers of multicultural America.

The world is constantly in flux and totally mixed and compounded. Nothing is really new. Creativity itself is a matter of seeing afresh what is already there and reading its implications and omens. (Stephen Owen's *Traditional Chinese Poetry and Poetics: Omen of the World* speaks to this point.) There are poems, novels, and paintings that roll onward through history, perennially redefining our places in the cosmos, that were initiated by such seeing. But creativity is not a unique, singular, godlike act of "making something." It is born of being deeply immersed in what *is*—and then seeing the overlooked connections, tensions, resonances, shadows, reversals, retellings. What comes forth is "new." This way of thinking about language is a world away from the usual ideas of education, however.

The standards of "Good Language Usage" until recently were based on the speech of people of power and position, whose language was that of the capital (London or Washington), and these standards were tied to the recognition of the social and economic advantages that accrue to their use. Another kind of standard involves a technical sort of writing that is dedicated to clarity and organization and is rightly perceived as an essential element in the tool kit of a person hoping for success in the modern world. This last sort of writing is intrinsically boring, but it has the usefulness of a tractor that will go straight and steady up one row and down another. Like a tractor, it is expected to produce a yield: scholarly essays and dissertations, grant proposals, charges or countercharges in legalistic disputes, final reports, long-range scenarios, strategic plans.

Truly Excellent Writing, however, comes to those who have learned, mastered, and passed through conventional Good Usage and Good Writing, and then loop back to the enjoyment and unencumbered playfulness of Natural Language. Ordinary Good Writing is like a garden that is producing exactly what you want, by virtue of lots of weeding and cultivating. What you get is what you plant, like a row of beans. But *really* good writing is both inside and outside the garden fence. It can be a few beans, but also some wild poppies, vetches, mariposa lilies, ceanothus, and some juncos and yellow jackets thrown in. It is more diverse, more interesting, more unpredictable, and engages with a much broader, deeper kind of intelligence. Its connection to the wildness of language and imagination helps give it power.

This is what Thoreau meant by the term "Tawny Grammar," as he wrote (in the essay "Walking") of "this vast, savage, howling mother of ours, Nature, lying all around, with such beauty, and such affection for her children, as the leopard; and yet we are so early weaned from her breast to society. . . . The Spaniards have a good term to express this wild and dusky knowledge, *Gramatica parda*, tawny grammar, a kind of mother-wit derived from that same leopard to which I have referred." The grammar not only of language, but of culture and civilization itself, comes from this vast mother of ours, nature. "Savage, howling" is another way of describing "graceful dancer" and "fine writer." (A linguist friend once commented, "Language is like a Mother Nature of feeling: it's so powerfully ordered there's room to be 99 percent wild.")

We can and must teach our young people to master the expected standard writing procedures, in preparation for the demands of multinational economies and of information overload. They will need these skills not only to advance in our postindustrial precollapse world, but also to critique and transform it. Those young learners with charming naive writing talents may suffer from the destructive effect of this discipline, because they will be brought to doubt their own ear and wit. They need to be assured that their unique personal visions will survive. They can take a deep breath and leap into the current formalities and rules, learn the game, and still come home to the language of heart and 'hood. We must continually remind people that language and its powers are far vaster than the territory deemed "proper usage" at any given time and place, and that there have always been geniuses of language who have created without formal education. Homer was a singer-storyteller, not a writer.

So, the more familiar view of language is:

1. Language is uniquely human and primarily cultural.
2. Intelligence is framed and developed by language.
3. The world is chaotic, but language organizes and civilizes it.
4. The more cultivated the language—the more educated and precise and clear—the better it will tame the unruly world of nature and feeling.
5. Good writing is "civilized" language.

But one can turn this around to say:

1. Language is basically biological; it becomes semicultural as it is learned and practiced.
2. Intelligence is framed and developed by all kinds of interactions with the world, including human communication, both linguistic and nonlinguistic; thus, language plays a strong—but not the only—rol⁻ in the refinement of thinking.

3. The world (and mind) is orderly in its own fashion, and linguistic order reflects and condenses that order.
4. The more completely the world is allowed to come forward and instruct us (without the interference of ego and opinion), the better we can see our place in the interconnected world of nature.
5. Good writing is "wild" language.

The twelfth-century Zen Buddhist philosopher Dōgen put it this way: "To advance your own experience onto the world of phenomena is delusion. When the world of phenomena comes forth and experiences itself, it is enlightenment." To see a wren in a bush, call it "wren," and go on walking is to have (self-importantly) seen nothing. To see a bird and stop, watch, feel, forget yourself for a moment, be in the bushy shadows, maybe then feel "wren"—that is to have joined in a larger moment with the world.

In the same way, when we are in the act of playful writing, the mind's eye is roaming, seeing sights and scenes, reliving events, hearing and dreaming at the same time. The mind may be reliving a past moment entirely in this moment, so that it is hard to say if the mind is in the past or in some other present. We move mentally as in a great landscape, and return from it with a few bones, nuts, or drupes, which we keep as language. We write to deeply heard but distant rhythms, out of a fruitful darkness, out of a moment without judgment or object. Language is a part of our body and woven into the seeing, feeling, touching, and dreaming of the whole mind as much as it comes from some localized "language center." Full of the senses, as

> Sabrina fair
>   Listen where thou art sitting
> Under the glassy, cool translucent wave,
>   In twisted braids of lilies knitting
> The loose train of thy amber-dropping hair;
>   Listen for dear honor's sake,
>   Goddess of the silver lake,
>   Listen and save.

Milton, Comus

Chill clarity, fluid goddess, silvery waves, and silvery flowers.

The faintly visible traces of the world are to be trusted. We do not need to organize so-called chaos. Discipline and freedom are not opposed to each other. We are made free by the training that enables us to master necessity, and we are made disciplined by our free choice to undertake mastery. We go beyond being a "master" of a situation by becoming a friend of "necessity" and thus—as Camus would have put it—neither victim nor executioner. Just a person playing in the field of the world.

[1995]

III

Watersheds

Reinhabitation

I came to the Pacific slope by a line of people that somehow worked their way west from the Atlantic over 150 years. One grandfather ended up in the Territory of Washington and homesteaded in Kitsap County. My mother's side were railroad people down in Texas, and before that they'd worked the silver mines in Leadville, Colorado. My grandfather being a homesteader and my father a native of the state of Washington put our family relatively early in the Northwest. But there were people already there, long before my family, I learned as a boy. An elderly Salish Indian gentleman came by our farm once every few months in a Model T truck, selling smoked salmon. "Who is he?" "He's an Indian," my parents said.

Looking at all the different trees and plants that made up my second-growth Douglas fir forest plus cow pasture childhood universe, I realized that my parents were short on a certain kind of knowledge. They could say, "That's a Doug fir, that's a cedar, that's bracken fern," but I perceived a subtlety and complexity in those woods that went far beyond a few names.

As a child I spoke with the old Salishan man a few times over the years he made these stops—then, suddenly, he never came back. I sensed what he represented, what he knew, and what it meant to me: he knew better than anyone else I had ever met *where I was*. I had no notion of a white American or European heritage providing an identity; I defined my-self by relation to the place. Later I also understood that "English language" is an identity—and later, via the hearsay of books, received the full cultural and historical view—but never forgot, or left, that first ground, the "where" of our "who are we?"

There are many people on the planet now who are not "inhabitants." Far from their home villages; removed from ancestral territories; moved into town from the farm; went to pan gold in California—work on the pipeline—work for Bechtel in Iran. Actual inhabitants—peasants, paisanos, paysan, peoples of the land, have been dismissed, laughed at, and overtaxed for centuries by the urban-based ruling elites. The intellectuals haven't the least notion of what kind of so-phisticated, attentive, creative intelligence it takes to "grow food." Virtually all the plants in the gardens and the trees in the orchards, the sheep, cows, and goats in the pastures were domesticated in the Neolithic, before "civilization." The dif-fering regions of the world have long had—each—their own precise subsistence pattern developed over millennia by people who had settled in there and learned what particular kinds of plants the ground would "say" at that spot.

Humankind also clearly wanders. Four million years ago

those smaller protohumans were moving in and out of the edges of forest and grassland in Africa—fairly warm, open enough to run in. At some point moving on, catching fire, sewing clothes, swinging around the arctic, setting out on amazing sea voyages. During the middle and late Pleistocene, large-fauna hunting era, a fairly nomadic grassland-and-tundra hunting life was established, with lots of mobility across northern Eurasia in particular. With the decline of the Ice Age—and here's where we are—most of the big-game hunters went out of business. There was possibly a population drop in Eurasia and the Americas, as the old techniques no longer worked.

Countless local ecosystem habitation styles emerged. People developed specific ways to *be* in each of those niches: plant knowledge, boats, dogs, traps, nets, fishing—the smaller animals and smaller tools. From steep jungle slopes of Southwest China to coral atolls to barren arctic deserts— *a spirit of what it was to be there* evolved that spoke of a direct sense of relation to the "land"—which really means, the totality of the local bioregion system, from cirrus clouds to leaf mold.

Inhabitory peoples sometimes say, "This piece of land is sacred"—or "all the land is sacred." This is an attitude that draws on awareness of the mystery of life and death, of taking life to live, of giving life back—not only to your own children but to the life of the whole land.

Abbé Breuil, the French prehistorian who worked extensively in the caves of southern France, has pointed out that the animal murals in those twenty-thousand-year-old caves describe fertility as well as hunting—the birth of little bison and cow calves. They show a tender and accurate observation of the qualities and personalities of different creatures, implying a sense of the mutuality of life and death in the food chain and what I take to be a sense of the sacramental quality of that relationship.

Inhabitation does not mean "not traveling." The term does not of itself define the size of a territory. The size is determined by the bioregion type. The bison hunters of the great plains are as surely in a "territory" as the Indians of northern California, though the latter may have seldom ventured farther than thirty miles from where they were born. Whether a vast grassland or a brushy mountain, the Peoples knew their geography. Any member of a hunting society could recall and visualize any spot in the surrounding landscape and tell you what was there, how to get there. "That's where you'd get some cattails." The bushmen of the Kalahari Desert could locate a buried ostrich egg full of emergency water in the midst of a sandy waste—walk right up and dig it out: "I put this here three years ago, just in case."

As always, Ray Dasmann's terms are useful to make these distinctions: "ecosystem-based cultures" and "biosphere cultures." By that Dasmann means societies whose life and economies are centered in terms of natural regions and watersheds, as against those who discovered—seven or eight thousand years ago in a few corners of the globe—that it was "profitable" to spill over into another drainage, another watershed, another people's territory, and steal away its resources, natural or human. Thus, the Roman Empire would strip whole provinces for the benefit of the capital, and villa-owning Roman aristocrats would have huge slave-operated farms in the south using giant wheeled plows. Southern Italy never recovered. We know the term *imperialism*—Dasmann's concept of "biosphere cultures" helps us realize that biological exploitation is a critical part of imperialism, too: the species made extinct, the clear-cut forests.

All that wealth and power pouring into a few centers had bizarre results. Philosophies and religions based on fascination with society, hierarchy, manipulation, and the "absolute." A great edifice called "the state" and the symbols of central power—in China what they used to call "the true dragon"; in the West, as Mumford says, symbolized perhaps by that Bronze Age fort called the Pentagon. No wonder Lévi-Strauss says that civilization has been in a long decline since the Neolithic.

So here in the twentieth century we find Occidentals and Orientals studying each other's wisdom, and a few people on both sides studying what came before both—before they forked off. A book like *Black Elk Speaks*, which would probably have had zero readership in 1900, is perceived now as speaking of certain things that nothing in the Judeo-Christian tradition, and very little in the Hindu-Buddhist tradition, deals with. All the world religions remain primarily human-centered. That next step is excluded or forgotten—"well, what do you say to Magpie? What do you say to Rattlesnake when you meet him?" What do we learn from Wren, and Hummingbird, and Pine Pollen, and how? Learn what? Specifics: how to spend a life facing the current; or what it is perpetually to die young; or how to be huge and calm and eat *anything* (Bear). But also, that we are many selves looking at each other, through the same eye.

The reason many of us want to make this step is simple, and is explained in terms of the forty-thousand-year looping back that we seem to be involved in. Sometime in the last twenty years the best brains of the Occident discovered to their amazement that we live in an Environment. This discovery has been forced on us by the realization that we are approaching the limits of something. Stewart Brand said that the photograph of the earth (taken from outer space by a satellite) that shows the whole blue orb with spirals and whorls of cloud was a great landmark for human consciousness. We see that it has a shape, and it has limits. We are back again, now, in the position of our Mesolithic forebears—working off the coasts of southern Britain, or the shores of Lake Chad, or the swamps of Southeast China, learning how to live by the sun and the green at that spot. We once more know that we live in a system that is enclosed in a certain way, that has its own kinds of limits, and that we are interdependent with it.

The ethics or morality of this is far more subtle than merely being nice to squirrels. The biological-ecological sciences have been laying out (implicitly) a spiritual dimension. We must find our way to seeing the mineral cycles, the water cycles, air cycles, nutrient cycles as sacramental—and we must incorporate that insight into our own personal spiritual quest and integrate it with all the wisdom teachings we have received from the nearer past. The expression of it is simple: feeling gratitude to it all; taking responsibility for your own acts; keeping contact with the sources of the energy that flow into your own life (namely dirt, water, flesh).

Another question is raised: is not the purpose of all this

living and studying the achievement of self-knowledge, self-realization? How does knowledge of place help us know the Self? The answer, simply put, is that we are all composite beings, not only physically but intellectually, whose sole individual identifying feature is a particular form or structure changing constantly in time. There is no "self" to be found in that, and yet oddly enough, there is. Part of you is out there waiting to come into you, and another part of you is behind you, and the "just this" of the ever-present moment holds all the transitory little selves in its mirror. The Avatamsaka ("Flower Wreath") jeweled-net-interpenetration-ecological-systems-emptiness-consciousness tells us no self-realization without the Whole Self, and the whole self is the whole thing.

Thus, knowing who we are and knowing where we are are intimately linked. There are no limits to the possibilities of the study of *who* and *where*, if you want to go "beyond limits"—and so, even in a world of biological limits, there is plenty of open mind-space to go out into.

Summing Up

In Wendell Berry's essay "The Unsettling of America," he points out that the way the economic system works now, you're penalized if you try to stay in one spot and do anything well. It's not just that the integrity of Native American land is threatened, or national forests and parks; it's *all* land that's under the gun, and any person or group of people who tries to stay there and do some one thing well, long enough to be able to say, "I really love and know this place," stands to be penalized. The economics of it works so that anyone who jumps at the chance for quick profit is rewarded—doing proper agriculture means *not* to jump at the most profitable chance—proper forest management or game management means doing things with the far future in mind—and the future is unable to pay us for it right now. Doing things right means living as though your grandchildren would also be alive, in this land, carrying on the work we're doing right now, with deepening delight.

I saw old farmers in Kentucky last spring who belong in another century. They are inhabitants; they see the world they know crumbling and evaporating before them in the face of a different logic that declares, "Everything you know, and do, and the way you do it, mean nothing to us." How much more the pain and loss of elegant cultural skills on the part of the nonwhite Fourth World primitive remnant cultures—who may know the special properties of a certain plant or how to communicate with dolphins, skills the industrial world might never regain. Not that special, intriguing knowledges are the real point: it's the sense of the magic system, the capacity to hear the song of Gaia *at that spot*, that's lost.

Reinhabitory refers to the tiny number of persons who come out of the industrial societies (having collected or squandered the fruits of eight thousand years of civilization) and then start to turn back to the land, back to place. This comes for some with the rational and scientific realization of interconnectedness and planetary limits. But the actual demands of a life committed to a place, and living somewhat by the sunshine green-plant energy that is concentrating in that spot, are so physically and intellectually intense that it is a moral and spiritual choice as well.

Mankind has a rendezvous with destiny in outer space, some have predicted. Well: we are already traveling in space—this is the galaxy, right here. The wisdom and skill of those who studied the universe firsthand, by direct knowledge and experience, for millennia, both inside and outside themselves, are what we might call the Old Ways. Those who envision a possible future planet on which we continue that study, and where we live by the green and the sun, have no choice but to bring whatever science, imagination, strength, and political finesse they have to the support of the inhabitory people—natives and peasants of the world. In making common cause with them, we become "reinhabitory." And we begin to learn a little of the Old Ways, which are outside of history, and forever new.

[This essay is based on a talk given at the Reinhabitation Conference at North San Juan School, held under the auspices of the California Council on the Humanities, August 1976. It was published in The Old Ways *(San Francisco: City Lights, 1977).]*

The Porous World

CRAWLING

I was forging along the crest of a ridge, finding a way between stocky deep red mature manzanita trunks, picking
the route and heading briskly on. Crawling.

Not hiking or sauntering or strolling, but *crawling*,
steady and determined, through the woods. We usually visualize an excursion into the wild as an exercise of walking
upright. We imagine ourselves striding through open alpine
terrain—or across the sublime space of a sagebrush basin—
or through the somber understory of an ancient sugar-pine
grove.

But it's not so easy to walk upright through the late-
twentieth-century midelevation Sierra forests. There are always many sectors regenerating from fire or logging, and the
fire history of the Sierra would indicate that there have always been some areas of manzanita fields. So people tend to
stay on the old logging roads or the trails, and this is their

way of experiencing the forest. Manzanita and ceanothus fields, or the brushy ground cover and understory parts of the forest, are left in wild peace.

This crawl was in late December, and although the sky was clear and sunny, the temperature was around freezing. Patches of remnant snow were on the ground. A few of us were out chasing corners and boundary lines on the Bear Tree parcel (number 6) of the 'Inimim Community Forest with a retiring Bureau of Land Management forester, a man who had worked with that land many years before and still remembered the surveys. No way to travel off the trail but to dive in: down on your hands and knees on the crunchy manzanita leaf cover and crawl around between the trunks. Leather work gloves, a tight-fitting hat, long-sleeved denim work jacket, and old Filson tin pants make a proper crawler's outfit. Along the ridge a ways, and then down a steep slope through the brush, belly-sliding on snow and leaves like an otter—you get limber at it. And you see the old stumps from early logging surrounded by thick manzanita, still-tough pitchy limbs from old wolf trees, hardy cones, overgrown drag roads, four-foot butt logs left behind, webs of old limbs and twigs and the periodic prize of a bear scat. So, face right in the snow, I came on the first of many bear tracks.

Later, one of our party called us back a bit: "A bear tree!" And sure enough, there was a cavity in a large old pine that had opened up after a fire had scarred it. A definite black bear hangout, with scratches on the bark. To go where bears, deer, raccoons, foxes—all our other neighbors—go, you have to be willing to crawl.

So we have begun to overcome our hominid pride and

learned to take pleasure in turning off the trail and going directly into the brush, to find the contours and creatures of the pathless part of the woods. Not really pathless, for there is the whole world of little animal trails that have their own logic. You go down, crawl swift along, spot an opening, stand and walk a few yards, and go down again. The trick is to have no attachment to standing; find your body at home on the ground, be a quadruped, or if necessary, a snake. You brush cool dew off a young fir with your face. The delicate aroma of leaf molds and mycelium rise from the tumbled humus under your hand, and a half-buried young boletus is disclosed. You can *smell* the fall mushrooms when crawling.

We began to fantasize on the broader possibilities of crawling. We could offer Workshops in Power Crawling! And in self-esteem—no joke! Carole said, "I've learned an important lesson. You can attain your goals, if you're willing to crawl!"

It's not always easy, and you can get lost. Last winter we took a long uphill cross-country transect on some of the land just above the Yuba Gorge; this soon turned into a serious crawl. We got into denser and denser old manzanita that had us doing commando-style lizard crawls to get under the very low limbs. It became an odd and unfamiliar ridge, and I had no idea where we might be. For hundreds of yards, it seemed, we were scuttling along, and then we came on a giant, totally fresh, worm-free *Leccinum manzanitae*, the prize of all the boletes. That went into the little day pack. And a bit farther the manzanita opened and there we were! We were in a gap below an old cabin built half onto BLM land at the edge of the Hindu yoga camp, and soon we found the dirt road that led toward home. One more victorious expedition through the underbrush.

As wide open spaces shrink around us, maybe we need to discover the close-up charms of the brushlands, and their little spiders, snakes, ticks (yikes!), little brown birds, lizards, wood rats, mushrooms, and poison-oak vines. It's not for everyone, this world of little scats and tiny tracks. But for those who are bold, I'd say get some gloves and a jacket and a hat and go out and *explore California*.

LIVING IN THE OPEN

One can choose to live in a place as a sort of visitor, or try to become an inhabitant. My family and I decided from early on to try to be here, in the midelevation forests of the Sierra Nevada, as fully as we could. This brave attempt was backed by lack of resources and a lot of dumb bravado. We figured that simplicity would of itself be beautiful, and we had our own extravagant notions of ecological morality. But necessity was the teacher that finally showed us how to live as part of the natural community.

It comes down to how one thinks about screens, fences, or dogs. These are often used for keeping the wild at bay. ("Keeping the wild at bay" sounds like fending off hawks and bears, but it is more often a matter of holding back carpenter ants and deer mice.) We came to live a permeable, porous life in our house set among the stands of oak and pine. Our buildings are entirely opened up for the long Sierra summer. Mud daubers make their trips back and forth from inside the house to the edge of the pond like tireless little cement trucks, and pour their foundations on beams, in cracks, and (if you're not alert) in rifle-bore holes and back-pack fire-pump nozzles. They dribble little spots of mud as they go. For mosquitoes, which are never much of a problem, the house is just another place to enjoy the shade. At night the bats dash around the rooms, in and out of the open skylights, swoop down past your cheek and go out an open sliding door. In the dark of the night the deer can be heard stretching for the lower leaves of the apple trees, and at dawn the wild turkeys are strolling a few yards from the bed.

The price we pay is the extra effort to put all the pantry food into jars or other mouse-proof containers. Winter bedding goes into mouse-proof chests. Then ground squirrels come right inside for fresh fruit on the table, and the deer step into the shade shelter to nibble a neglected salad. You are called to a hopeful steadiness of nerves as you lift a morsel of chicken to the mouth with four meat bees following it every inch of the way. You must sometimes (in late summer) cook and eat with the yellow jackets watching every move. This can make you peevish, but there is a kind of truce that is usually attained when one quits flailing and slapping at the wasps and bees.

It's true, living and cooking in the outdoor shade shelters someone occasionally gets stung. That's one price you might pay for living in the porous world, but it's about the worst that can happen. There's a faint risk of rattlesnake bite as we stride around the little trails, and the ever-present standoff-ishness of poison oak. But if you can get used to life in the semiopen, it's a great way to enjoy the forest.

It's also a form of conservation. As people increasingly come to inhabit the edges and inholdings of forest lands, they have to think carefully about how they will alter this new-old habitat. The number of people that can be wisely accommodated on the land cannot be determined simply by saying how many acres are required for a single household. This kind of planning is essential right now, and I'm all for it, but we have to remember that the cultural practices of households alone can make huge differences in impact.

Necessary roads should be thoughtfully routed and of modest width, with the occasional fire-truck turnout. Fire protections should be provided by having the roads well brushed along the edges, with plenty of thinning back into the woods, rather than building an excessively wide road-bed. If roads are a bit rough, it will slow cars down, and that's not all bad. If there are no or very few fences, if people are not pumping too heavily from their wells to irrigate pasture or orchards, if the number of dogs is kept modest, if the houses are well insulated and temperatures are held at the low sixties in the winter, if feral cats are not allowed, if an attitude of tolerance is cultivated toward the occasional mischief of critters, we will cause almost no impact on the larger forest ecosystem. But if there are too many people who hate insects and coyotes, who are perpetually annoyed by deer and who get hysterical about bears and cougars, there goes the neighborhood.

It's possible and desirable to take out firewood lightly, to cut some deliberately chosen sawlogs, to gather manzanita berries for the cider, to seek redbud for basketry supplies, and to pursue any of a number of other subtle economic uses of the forest. As we thin saplings, remove underbrush, and move tentatively toward the occasional prescribed burn, we are even helping the forest go its own direction. Maybe we will yet find ways to go past the dichotomy of the wild and the cultivated. Coyotes and screech owls make the night magic; log-truck airhorns are an early morning wakeup.

Permeability, porousness, works both ways. You are allowed to move through the woods with new eyes and ears when you let go of your little annoyances and anxieties. Maybe this is what the great Buddhist philosophy of interconnectedness means when it talks of "things moving about in the midst of each other without bumping."

["Living in the Open" (1991) and "Crawling" (1992) appeared in numbers 2 and 3 of Tree Rings, *the newsletter of the Yuba Watershed Institute.]*

The Forest in the Library

I prepared a talk for the October 19, 1990, dedication of the new West Wing of Shields Library, University of California at Davis.

*I*n the old and original spirit of dedications, and in honor of the life of buildings, I want to invoke the many presences that are here—not invisible, just rarely seen—whose goodwill toward this project can certainly be hoped for. We are right on the territory of the old Patwin village of Putah-toi, which was a large, settled, and affluent community whose memories went back several thousand years. May the deeply conservative spirit of the Native Californians, and their love for lore and the rituals that preserve it, welcome this structure to a long and useful life. May the even older presences here—the valley oaks and in particular the great oak within the courtyard (bemused as it may be by recent changes), the Swainson's hawks that soar past the top of Sproul Hall, the burrowing owls, and Putah Creek itself (reduced as it is for the moment)—lend their support to this current human effort of a university and a library. May the trees that were sacrificed for this expansion be justified by the good work that should come forth. We devoutly hope that this large enterprise will serve the welfare of watersheds, owls, trees, and, of course, human beings.

As for this new wing itself, it is an elegant structure of cast-in-place concrete—that is to say, a transformation of water-washed gravels, a riverbed stood on end. The architects tell me that this new part of the building is substantially made up of old riverbeds of the Stanislaus River drainage, which has thus come over here visiting. We are, so to speak, now introducing these assembled elements to each other, that they may wish each other well.

It is also the case that in fin-de-millennium California we have much longer threads of connection: in addition to the historical links eastward to Europe and Africa, we now look westward to Polynesia and Asia in matters both ecological and economic. We have historical and cultural connections to the south with Hispanic culture, and the great Pacific flyway brings the Canada geese and pintail ducks from their nesting grounds in the far north to the marshes just beyond the campus. All of these lineages are present in our daily lives and are literally represented in the cosmopolitanism of our student body and the diversity of our studies. This is all to be welcomed, even as we simultaneously celebrate the antiquity and resilience of the original nature of our treasured California landscape.

We live at the intersection of many forces, and in the case of the library in particular, there is one more force to be invoked. That is our occidental humanistic and scientific intellectual tradition. It has demonstrated an extraordinary ability to maintain itself through time. The institution of the library is at the heart of that persistence. Although Strabo said, "Aristotle was the first man to have collected books," there were in truth hundreds of outstanding private libraries in Hellenic Greece. What survived of Aristotle's personal library became the basis of one of the first institutional libraries, which soon became a feature of classical civilization. There were, of course, far older libraries, and in the broader sense archives of literature and lore were kept worldwide, in virtually all cultures whether they had writing or not.

The original context of teaching must have been narratives told by elders to young people gathered around the fire. Our fascination with TV may just be nostalgia for that flickering light. My grandparents didn't tell stories around the campfire before we went to sleep—their house had an oil furnace instead, and a small collection of books. I got into their little library to entertain myself. In this huge old occidental culture, our teaching elders are books. For many of us, books are our grandparents! In the library there are useful, demanding, and friendly elders available to us. I like to think of people like Bartolomé de las Casas, who passionately defended the Indians of New Spain, or Baruch Spinoza, who defied the traditions of Amsterdam to be a philosopher. (And in my days as an itinerant forest worker I made especially good use of libraries: they were warm and stayed open late at night.)

Making hoards and heaps, saving lore and information, are entirely natural: some zooarchaeologists have excavated heaped-up wood-rat nests out in the Mojave Desert, packed full of little wood-rat treasures, that are twelve thousand years old. We humans are truly just beginners.

Pursuing this line of thought, my friend Jack Hicks of the English Department and I were talking about how one might see the university as a natural system, and wondering what the information flow would look like. We found ourselves, in this year of forest consciousness, recalling the venerable linkage of academies to groves. In China, too, academies such as the Han-lin were called "groves." We considered that the information web of the modern institution of learning, right down to the habitat niches of buildings, has an energy flow fueled by the data accumulation of primary workers in the information chain—namely the graduate students and young scholars. Some are green like grass, basic photosynthesizers, grazing brand-new material. Others are in the detritus cycle and are tunneling through the huge logs of old science and philosophy and literature left on the ground by the past, breaking them down with deconstructive fungal webs and converting them anew to an edible form. These people on the floor of the information forest are among the hardest workers, and to be sure are affrighted occasionally by hawklike shadows sailing over them.

The gathered nutrients are stored in a place called the *bibliotek*, "place of the papyrus," or the *library*, "place of bark," because the Latin word for tree bark and book is the same, reflecting the memory of the earliest fiber used for writing in that part of the Mediterranean.

If you will allow me to carry this playful ecological analogy a little further, we can say that the dissertations, technical reports, and papers of the primary workers are in a sense gobbled up by senior researchers and condensed into conclusion and theory—new studies that are in turn passed up the information chain to the thinkers at the top who will digest them and come out with some unified theory or perhaps a new paradigm. These final texts, which are built on the concentrated information assembled lower on the chain, will be seen as the noble monarchs of the academy-forest. Such giants must also succumb in time and return to the forest floor.

When asked "What is finally over the top of all the information chains?" one might reply that it must be the artists and writers, because they are among the most ruthless and efficient information predators. They are light and mobile, and can swoop across the tops of all the disciplines to make off with what they take to be the best parts, and convert them into novels, mythologies, dense and esoteric essays, visual or other arts, or poems. And who eats the artists and writers? The answer must be that they are ultimately recycled to the beginners, the students. That's where the artists and writers go, to be cheerfully nibbled and passed about.

The library itself is the heart of this ancient forest. But as Robert Gordon Sproul [former president of the University of California] said in his highly regarded speech of 1930, the library would be useless just as a simple collection of books or information. It is the organization, the intelligent system that can swiftly seek out and present one tiny bit of its stored information to a single person, that makes it useful. What lies behind it all, of course, is language. As I have written elsewhere, language is a mind-body system that coevolved with our needs and nerves. Like imagination and the body, language rises unbidden. It is of a complexity that eludes our rational intellectual capacities, yet the child learns the mother tongue early and has virtually mastered it by six. . . . Without conscious device we constantly reach into the vast word hoards in the depths of the wild unconscious. We cannot as individuals or even as a species take credit for this power; it came from someplace else, from the way clouds divide and mingle, from the way the many flowerlets of a composite blossom divide and redivide.

Yet acknowledging all that freshness and order from within, our inherent intellectual infrastructure, should only intensify our regard for the amazing *deliberateness* that has given us our institutions of higher learning, within which the library is another sort of relatively unappreciated infrastructure not unlike language itself. The refinement of organization makes a library work, and like the rich syntax of a natural language, it almost eludes us. For most of us, it borders on mystery and calls, if not for offerings, at least for gratitude. So I want to express the gratitude we must all feel for the good luck that has brought us together today, with this fine library, admiring its handsome newly extended shell, which will be serving the great project of world intellectual culture. We celebrate a new opening, a new step, in this old-new project of human self-knowledge.

[1990]

Exhortations for Baby Tigers

THE END OF THE COLD WAR
AND THE "END OF NATURE"

In May of 1991 I was invited to speak at the Reed College graduation ceremony. It was exactly forty years since my own graduation from the same institution.

*H*ere we are, on a rise of ground east of the north-flowing Willamette River, on a parcel straddling a small, remarkably cold little spring-fed creek. People have lived along the Willamette for millennia. This site is watched over by some Douglas fir trees that were witness to the first construction of the Reed buildings. They are the immediate descendants of the dense coniferous forest that once covered the western slopes of this continent, at the shore of the eastern Pacific. I suppose their million years of experience on this spot has amounted to a sort of "purification of information"; maybe they are all Doctors of Perennial Habitat, experts in staying put—and we are little grazers and grifters who are temporarily shouldering them aside. But we are also coming to be the people of the continent, as long as it takes. A salute, then, to the soils and trees and waters here that support this fine college, and to the memory of the hundreds of generations of refined and self-sufficient people who lived in the lower Willamette Valley.

It is clear that this generation, your generation, your five-year cohort, is "commencing"—beginning to get out there—at a truly pivotal time in history. Recent history has brought us up against two absolutely outstanding challenges: they are the end of the cold war, and the end of nature. One must be approached with courage and generosity, the other with what I would call "trans-species erotics."

The end of the cold war. For over a century now, people all over the world have been engaged in the struggle between revolutionary socialism (in its various forms and schools) and the free-market capitalist ideologies. The political confrontations have been violent and bitter. Still, within the hearts of the passionate advocates on both sides, the adversarial relationship has never been totally black and white. Business leaders of conscience have always been aware of the flaws and contradictions of the capitalist system, and countless socialists have been alarmed by the tendencies toward overcentralization, bureaucracy, and totalitarianism they could see inherent in their own "actually existing socialism." It was always one hopeful sort of imperfection set against another. But for now "capitalism" seems to have won the joust, and for the moment there is but one superpower on earth, the United States.

How tempting it will be for some North American leaders to look on this as an opportunity to have "one world run by one world power"—total hegemony—the complete Roman Empire of the twenty-first century. It would be launched under the rationalization of some need for a "Pax Americana." What a nightmare. Not because the United States is particularly evil, but because there should *never* be one world power. We need planetary diversity in nations as much as we need human diversity in society or biological diversity in the forest. So here's the first challenge: the question of who and what will now critique the rising tide of economic growth and consumerism and petty competitive nationalism that could lead the world into a kind of resource extraction race and to ecological ruin. A rush that would result in, as some wag put it, "energy for a brief America" and "strength through exhaustion."

For decades the socialists, with their theoretical emphasis on economic justice, mounted critiques of the seemingly heartless and heedless dynamics of capitalism. Now such critiques will have to come from within the "free-world" societies, or from whatever post-Marxist creative rebuilding of socialist praxis is possible in the future reconstitution of the former Soviet bloc. The capitalist world must rediscover its own conscience, now that it has no adversary. It's clear that the Americans of the U.S., the Japanese, and the western Europeans must seek some social and ecological positions that go beyond business, markets, and profits—or risk becoming what the Buddhists call perpetual Hungry Ghosts: creatures with enormous bellies, insatiable appetites, and tiny mouths.

Entering the market and career world now, you will be the ones called on to seek a new philosophy and practice of power here in the United States. What values your new view and practice will ultimately be grounded on is hard to guess. Rational long-range self-interest wouldn't be half-bad, for starters. Long-range self-interest would realize, for example, that deforestation today does not create jobs for the future.

The emphasis on human rights that is rooted in the Judeo-Christian tradition, the concern for all beings expressed in Buddhism, and the compassionate political savvy of Confucianism (which is responsible for much of the success of modern Japan) will contribute to it. The heritage of almost three centuries of European socialist and communal thought and speculation still provides lessons and possibilities. My own sympathies still lie with a kind of anarcho-syndicalist organization of work groups blocked out in terms of bioregions. *Courage* and *generosity* are key words in the practice of social values meant to serve society and not the state. Courage because, as Samuel Johnson quipped, none of the other virtues can be practiced without it, and generosity because this is what it takes to let the world go its own way, without needing to exploit or dominate. A Zen proverb suggests a kind of elegant unarmed fearlessness as an approach to this:

Walk in the dark
In your best clothes.

The second (and related) challenge, "the end of nature"—to borrow the title of Bill McKibben's book on
global warming—does not mean literally the end of physical nature but the end of the way we have been looking at nature, the end of our "construction" of nature for the past five
hundred years of occidental culture, whether feudal, mer-
cantile, socialist, or capitalist.

For me this means the end of taking the natural world for
granted as a kind of hardware store and lumberyard, to be
used and exploited to the maximum—a realm with no intrinsic value of its own. In the course of our coming to understand the interconnectedness of life and the remarkable
ways that energy flows through living systems on the planet,
we are possibly finally seeing that the time has come to put
aside unexamined human assumptions of species superiority and all the destruction that goes with them. We humans
might just learn to see ourselves as fully part of the transhu-
man realm. This means allowing the intrinsic *value*—soul,
if you like—of the rest of the world. There is nothing in our
whole occidental tradition to prepare us for such an attitude,
yet it is essential.

I recently read a well-intentioned essay in the humanist
journal *The American Prospect*, winter 1991 issue, by Alan
Wolfe. It is called "Up from Humanism" and is a sympa-
thetic review of recent animal rights and deep-ecology literature. The author proposes that our "human specialness"
lies in what he calls "our ability to attribute meaning to the

world around us." I would argue that he has it backward. Humans, and all the other organisms, have their own specialness—but it is the *world* that gives meaning to each of us. The multitudinous and various phenomena of the world sweeping over us every day teach us who we are. A Chinese poem-riddle puts it something like this:

> The forested ridges, the blossoming pears, the shifting
>   clouds—
> Who is it all for?

Part of the answer can be that the beautiful world of nature is for this marvelous sensitive human creature we are, but *also* and *equally* for the sleepy baby bat hanging in the eaves or the hummingbird on a courtship dive. Whatever sense of ethical responsibility and concern that human beings can muster must be translated from a human-centered consciousness to a natural-systems-wide sense of value. First, simply because such a bighearted sense of the world seems right, but also to help avert the potential destruction of even the very processes that sustain most life on earth. (The possibility of human agency causing some irreversible harm, such as global warming as described in McKibben's study, is quite real.)

Such an extension of human intellect and sympathy into the nonhuman realms is a charming and mind-bending undertaking. It is also an essential step if we are to have a future worth living. It was hinted at in our ancient past, and could, if accomplished, be the culminating human moral and aesthetic achievement. I imagine this possibility as a kind of "trans-species erotics." The worldwide myths of animal-human marriage, or supernatural-human marriage, are evidence of the fascination our ancestors had for the possibility of full membership in a biotic erotic universe. I suspect that many of the problems within the human community—racism and sexism, to name two—reflect back from confusion about our relation to nature. Ignorance and hostility toward wild nature set us up for objectifying and exploiting fellow humans.

The year I graduated from Reed College, 1951, the commencement speech was given by our keen, sardonic Professor Dick Jones. It was concerned with—as I recall—the decline of the West, which at that time sounded like a good idea. What has happened since the fifties has been far more complex than any particular cultural decline. America, for example, has become a rather nervous, prickly old country getting set in its ways, sensitive to slights and intolerant of dissent, while Japan and Taiwan are enthusiastically youthful and entrepreneurial societies with little taste for military rhetoric. Citizens of the Soviet Union and the eastern European nations have lost all faith in official ideologies, and many are putting their hope in the environmental movement and in the idea of smaller communities. We must devoutly wish that such hopes as these take precedence over the resurgence of old rivalries. The problems of Africa and Europe have been to a great degree created by the cynical drawing and redrawing of national boundaries by the European powers.

When I was a youth, there was a lot of talk about soil erosion and a certain concern for old-growth forest, but nothing was said about their seldom-seen owls. The fate of old-growth forests—and everything else—has become part of a world debate. It is my own sort of crankiness to believe there is still hope. I would like to think the technological society—as diverse, smart, and complicated as it is—can also get "nature literate" and be fully at home with the wild, both within and without.

So, for starters I propose that we take note of the locals as well as the universals, pay as much attention to the community as we do to the national state, keep looking for the questions rather than the answers, walk more and drive less, and follow our own integral concepts of whatever a good and productive life might be, rather than some mainstream image. As Robinson Jeffers wrote, "corruption/never has been compulsory." Some of the finest people who ever graduated from this or that university did not become famously successful. Some I know quite well, and they have been lost to their alumni associations for decades, while living exemplary lives and accomplishing needed and innovative work in the everyday world. They are, as a Taoist once put it, "sages disguised as melon growers in the mountains."

We should be dubious of fantasies that would lead toward centralizing world political power, but we do need to nourish interactive playful diversity on this one-planet watershed. And there's a Chinese Buddhist saying to *this* point:

> Easy to reach nirvana,
> Hard to enter difference.

May the Beaver and Douglas Fir Powers bless us (they do seem both to be totems of the West Coast, and of Reed). As my motorcycle buddies say, "ride hard, die free." Let's go on into the twenty-first century lean, mean, and green. One more Chinese proverb to leave you with, as you step out into the arena:

> A baby tiger just born on the ground
> Has an ox-eating spirit.

[1991]

Walt Whitman's Old "New World"

I honor Walt Whitman's marvelous exuberant vision of a "new world," particularly as brought forward in his poetry and the essay "Democratic Vistas." In its own time and place, it is a vibrant and inspiring hope for a world people yet to come, a society that would become the best on earth.

That new world proves to be its own kind of fiction. I am not talking about the fictions of idealized democracy as set against the disappointments of the twentieth century. There is something far more fundamental at stake. For the actual locus of the American "new world" is not new but is an ancient continent, hundreds of millions of years old. At the time of Columbus's voyage, the Western Hemisphere already had something like sixty million lively, totally competant, well-settled inhabitants.

Whitman's main focus is the vigorous daily life of

nineteenth-century American society. For him the "States" is a world of visionary potentiality. Even as he lashes out at the materialism and corrupt politics of his time, he also quite accurately predicts that America will be the leading nation of the future with populations in the hundreds of millions. He sees the future global communication system that electricity will make possible. The America that Walt Whitman envisions will have some excellent features. He suggests that the "strong and sweet female race" will be fully equal to men, with properly granted dignity—and (for him most important) the people of the future will have, will require, a literature that is truly spiritual, religious, and at the same time popular. But from the vantage point of the late twentieth century, as attractive as Whitman's prospect is, we can see that there are some key elements missing. In "Democratic Vistas" we miss the presence of people of color, of Native Americans, of wilderness, or even the plain landscape.

At the opening of the essay Whitman invokes the great lessons of nature, which are, he says, "variety" and "freedom." Take variety. (Today a common term for the natural variety of plant and animal species is *biodiversity*.) The science of conservation biology, a strong new field in the universities and one that plays a role in public policy questions, tells us that if we would maintain natural variety in North America we must keep millions of acres of land in wild or semiwild condition to provide the living space—marsh, grassland, or forest—required by nonhuman species. Oddly enough, there is little grasp of the issues concerning wild nature in Whitman's rhetoric. He celebrates industry, workers, action, and the relentless energy of Americans (who almost always are implicitly white). In his essay he finds nature either "healthy or happy" or finally "nothing in itself" and "serviceable." Whitman is unexcelled in his attribution of a kind of divinity to ordinary (white) men and women. However, the respect and authenticity he gives to human beings is not extended to nonhuman creatures. Yet North America was and is home to an extraordinary natural community. Fifty million bison and approximately twenty million pronghorn were on the Great Plains as late as the mid-nineteenth century. This was the largest single population of big mammals anywhere on earth.

Take human variety and freedom. Whitman was clearly free of prejudice in his personal manners and values. But he assumed a kind of melting-pot future in which the other races and ethnic groups would eventually become one with the liberal Protestant metaphysic that lurks behind his dream. We know today that this will not happen as hoped. The actual ideology of Whitman's projected future did not truly respect variety, nor could it know that different cultures would stubbornly remain, if they chose, *different*. Walt Whitman suggests that he can "embrace all, reject none." But the etiquette of mutual respect, both among human beings and between the human and nonhuman realms, is more tricky than this optimistic rhetoric suggests. As a vernacular community-based ecologist, I would suggest that we must imaginatively transform democracy into a trans-species exercise, not merely an in-house human-species political practice.

Whitman is rather like the "younger son" of fairy tale— footloose, cheerful, trusting, "adhesive," loosely and sometimes passionately spiritual, but in a certain sense always un-grounded, and always a *puer*. I love his life, his character, and his poetry. But it troubles me that the years during which Whitman wrote "Democratic Vistas" (1868–70) were years of defeat and misery for Native Americans, and were the very years when the commercial destruction of the North American bison herd was fully under way. Fifteen to twenty thousand buffalo hunters were killing and wasting literally millions of bison every season. It was not a time of variety and freedom for the native, or for nature.

There is a further vision beginning to take hold in North America now, which owes something to Whitman but which goes both forward and backward in time. The Native Americans of 1992, for one thing, are not about to accept merely being "embraced" by the ongoing obtuseness of even the most liberal Euro-Americans. The situation of the in-digenous people of North America, with its ineluctable power and pathos, has yet to be truly acknowledged in the United States. To my mind the most potent contemporary Native American thinking has gone beyond old political ar-guments over treaty rights and so on with the U.S. or Canada (though these arguments must still be addressed). The new thinking looks toward the emergence of the natural nations of the future. (An "Inuit nation" is in the making right now, for example, across northern Canada, binding the Inuit [Es-kimo] people of Alaska, Canada, Greenland, and Siberia into one.) The bioregional movement also calls for "natural nations" on Turtle Island (North America). This ecological/poetical exercise starts with an analysis of how the political boundaries of the American states—those between Canada and Mexico—can obscure the biological and cultural realities of the landscape.

It is also a new thought that anyone of any cultural or racial background who chooses to learn, love, and respect the North American continent and its human and nonhuman inhabitants—and its ecosystems and watersheds—can be a sort of honorary Native American. Rather than trying to build a new world, some of us celebrate the possibility of membership in the ancient world of Turtle Island. If Walt Whitman were with us today, we know he would be bitterly disappointed in the spiritual and cultural poverty of this America that became a global leader and the richest and most powerful nation on earth. He might well give his heart to the new native and bioregional movement with as much foolish optimism as he gave to his own uniquely enlightened version of the nineteenth-century fantasy of progress.

[This essay is based on a talk given at the Casa de América in Madrid, Spain, on December 1, 1992, for the "Homenaje a Walt Whitman en el Centenario de Su Muerte" (Homage to Walt Whitman on the Centenniul of His Death)]

Coming into the Watershed

I had been too long in the calm Sierra pine groves and wanted to hear surf and the cries of seabirds. My son Gen and I took off one February day to visit friends on the north coast. We drove out of the Yuba River canyon, and went north from Marysville—entering that soulful winter depth of pearly tule fog—running alongside the Feather River and then crossing the Sacramento River at Red Bluff. From Red Bluff north the fog began to shred, and by Redding we had left it behind. As we crossed the mountains westward from Redding on Highway 299, we paid special attention to the transformations of the landscape and trees, watching to see where the zones would change and the natural boundaries could be roughly determined. From the Great Valley with its tules, grasses, valley oak, and blue oak, we swiftly climbed into the steep and dissected Klamath range with its ponderosa pine, black oak, and manzanita fields. Somewhere past Burnt Ranch we were in the redwood and Douglas fir forests—soon it was the coastal range. Then we were descending past Blue Lake to come out at Arcata.

We drove on north. Just ten or fifteen miles from Arcata, around Trinidad Head, the feel of the landscape subtly changed again—much the same trees, but no open meadows, and a different light. At Crescent City we asked friends just what the change between Arcata and Crescent City was. They both said (to distill a long discussion), "You leave 'California.' Right around Trinidad Head you cross into the maritime Pacific Northwest." But the Oregon border (where we are expected to think "the Northwest" begins) is still many miles farther on.

So we had gone in that one afternoon's drive from the Mediterranean-type Sacramento Valley and its many plant alliances with the Mexican south, over the interior range with its dry pine-forest hills, into a uniquely Californian set of redwood forests, and on into the maritime Pacific Northwest: the edges of four major areas. These boundaries are not hard and clear, though. They are porous, permeable, arguable. They are boundaries of climates, plant communities, soil types, styles of life. They change over the millennia, moving a few hundred miles this way or that. A thin line drawn on a map would not do them justice. Yet these are the markers of the natural nations of our planet, and they establish real territories with real differences to which our economies and our clothing must adapt.

On the way back we stopped at Trinidad Head for a hike and a little birding. Although we knew they wouldn't be there until April, we walked out to take a look at the cliffs on the head, where tufted puffins nest. For tufted puffins, this is virtually the southernmost end of their range. Their more usual nesting ground is from southeastern Alaska through the Bering Sea and down to northern Japan. In winter they are far out in the open seas of the North Pacific. At this spot, Trinidad, we could not help but feel that we touched on the life realm of the whole North Pacific and Alaska. We spent that whole weekend enjoying "liminality," dancing on the brink of the continent.

I have taken to watching the subtle changes of plants and climates as I travel over the West. We can all tell stories, I know, of the drastic changes we have noticed as we raged over this or that freeway. This vast area called "California" is large enough to be beyond any one individual's ability (not to mention time) to travel over and to take it all into the imagination and hold it clearly enough in mind to see the whole picture. Michael Barbour, a botanist and lead author of *California's Changing Landscapes*, writes of the complexity of California: "Of the world's ten major soils, California has all ten. . . . As many as 375 distinctive natural communities have been recognized in the state. . . . California has more than five thousand kinds of native ferns, conifers, and flowering plants. Japan has far fewer species with a similar area. Even with four times California's area, Alaska does not match California's plant diversity, and neither does all of the central and northeastern United States and adjacent Canada combined. Moreover, about 30 percent of California's native plants are found nowhere else in the world."

But all this talk of the diversity of California is a trifle misleading. Of what place are we speaking? What is "California"? It is, after all, a recent human invention with hasty straight-line boundaries that were drawn with a ruler on a map and rushed off to an office in D.C. This is another illustration of Robert Frost's lines, "The land was ours before we were the land's." The political boundaries of the western states were established in haste and ignorance. Landscapes have their own shapes and structures, centers and edges, which must be respected. If a relationship to a place is like a marriage, then the Yankee establishment of a jurisdiction called California was like a shotgun wedding with six sisters taken as one wife.

California is made up of what I take to be about six regions. They are of respectable size and native beauty, each with its own makeup, its own mix of birdcalls and plant smells. Each of these proposes a slightly different lifestyle to the human beings who live there. Each led to different sorts of rural economies, for the regional differences translate into things like raisin grapes, wet rice, timber, cattle pasture, and so forth.

The central coast with its little river valleys, beach dunes and marshes, and oak-grass-pine mountains is one region. The great Central Valley is a second, once dominated by swamps and wide shallow lakes and sweeps of valley oaks following the streams. The long mountain ranges of the Sierra Nevada are a third. From a sort of Sonoran chaparral they rise to arctic tundra. In the middle elevations they have some of the finest mixed conifer forests in the world. The Modoc plateau and volcano country—with its sagebrush and juniper—makes a fourth. Some of the Sacramento waters rise here. The fifth is the northern coast with its deep interior mountains—the Klamath region—reaching (on the coast) as far north as Trinidad Head. The sixth (of these six sisters) consists of the coastal valleys and mountains south of the Tehachapis, with natural connections on into Baja. Although today this region supports a huge population with water drawn from the Colorado River, the Owens Valley, and the great Central Valley, it was originally almost a desert.

One might ask, What about the rest? Where are the White Mountains, the Mojave Desert, the Warner Range? They are splendid places, but they do not belong with California. Their watersheds and biological communities belong to the Great Basin or the lower Colorado drainage, and we should let them return to their own families. Almost all of core California has a summer-dry Mediterranean climate, with (usually) a fairly abundant winter rain. More than anything else, this rather special type of climate is what gives our place its fragrance of oily aromatic herbs, its olive-green drought-resistant shrubs, and its patterns of rolling grass and dark forest.

I am not arguing that we should instantly redraw the boundaries of the social construction called California, although that could happen some far day. But we are becoming aware of certain long-range realities, and this thinking leads toward the next step in the evolution of human citizenship on the North American continent. The usual focus of attention for most Americans is the human society itself with its problems and its successes, its icons and symbols. With the exception of most Native Americans and a few non-natives who have given their hearts to the place, the land we all live on is simply taken for granted—and proper relation to it is not considered a part of "citizenship." But after two centuries of national history, people are beginning to wake up and notice that the United States is located on a landscape with a severe, spectacular, spacy, wildly demanding, and ecstatic narrative to be learned. Its natural communities are each unique, and each of us, whether we like it or not—in the city or countryside—lives in one of them.

Those who work in resource management are accustomed to looking at many different maps of the landscape. Each addresses its own set of meanings. If we look at land ownership categories, we get (in addition to private land) the Bureau of Land Management, national forest, national park, state park, military reserves, and a host of other public holdings. This is the public domain, a practice coming down from the historic institution of the commons in Europe. These lands, particularly in the arid West, hold much of the water, forest, and wildlife that are left in America. Although they are in the care of all the people, they have too often been managed with a bent toward the mining or logging interests and toward short-term profits.

Conservationists have been working since the 1930s for sustainable forestry practices and the preservation of key blocks of public land as wilderness. They have had some splendid success in this effort, and we are all indebted to the single-minded dedication of the people who are behind every present-day wilderness area that we and our children walk into. Our growing understanding of how natural systems work brought us the realization that an exclusive emphasis on disparate parcels of land ignored the insouciant freeness of wild creatures. Although individual islands of wild land serving as biological refuges are invaluable, they cannot by themselves guarantee the maintenance of natural variety. As biologists, public land managers, and the involved public have all agreed, we need to know more about how the larger-scale natural systems work, and we need to find "on-the-ground" ways to connect wild zone to wild zone wherever possible. We have now developed the notion of biological corridors or connectors. The Greater Yellowstone Ecosystem concept came out of this sort of recognition. Our understanding of nature has been radically altered by systems theory as applied to ecology, and in particular to the very cogent subdisciplines called island biogeography theory and landscape ecology.

No single group or agency could keep track of grizzly bears, which do not care about park or ranch boundaries and have necessary, ancient territories of their own that range from late-summer alpine huckleberry fields to lower-elevation grasslands. Habitat flows across both private and public land. We must find a way to work with wild ecosystems that respects both the rights of landowners and the rights of bears. The idea of ecosystem management, all the talk now in land management circles, seems to go in the right direction. Successfully managing for the ecosystem will require as much finesse in dealing with miners, ranchers, and motel owners as it does with wild animals or bark beetles.

A "greater ecosystem" has its own functional and structural coherence. It often might contain or be within a watershed system. It would usually be larger than a county, but smaller than a western U.S. state. One of the names for such a space is "bioregion."

A group of California-based federal and state land managers who are trying to work together on biodiversity problems recently realized that their work could be better accomplished in a framework of natural regions. Their interagency "memorandum of understanding" calls for us to "move beyond existing efforts focused on the conservation of individual sites, species, and resources . . . to also protect and manage ecosystems, biological communities, and landscapes." The memorandum goes on to say that "public agencies and private groups must coordinate resource management and environmental protection activities, emphasizing regional solutions to regional issues and needs."

The group identified eleven or so such working regions within California, making the San Francisco Bay and delta into one, and dividing both the Sierra and the valley into northern and southern portions. (In landscapes as in taxonomy, there are lumpers and splitters.) Since almost 50 percent of California is public domain, it is logical that the chiefs of the BLM, the Forest Service, California Department of Fish and Game, California Department of Forestry, State Parks, the federal Fish and Wildlife Service, and such should take these issues on, but that they came together in so timely a manner and signed onto such a far-reaching plan is admirable.

Hearing of this agreement, some county government people, elected officials, and timber and business interests in the mountain counties went into a severe paranoid spasm, fearing—they said—new regulations and more centralized government. So later in the fall, an anonymous circular made its way around towns and campuses in northern California under the title "Biodiversity or New Paganism?" It says that "California Resource Secretary Doug Wheeler and his self-appointed bioregional soldiers are out to devalue human life by placing greater emphasis on rocks, trees, fish, plants, and wildlife." It quotes me as having written that "those of us who are now promoting a bioregional consciousness would, as an ultimate and long-range goal, like to see this continent more sensitively redefined, and the natural regions of North America—Turtle Island—gradually begin to shape the political entities within which we work. It would be a small step toward the deconstruction of America as a superpower into seven or eight natural nations—none of which have a budget big enough to support missiles." I'm pleased to say I did write that. I'd think it was clear that my statement is not promoting more centralized government, which seems to be a major fear, but these gents want both their small-town autonomy and the military-industrial state at the same time. Many a would-be westerner is a rugged individualist in rhetoric only, and will scream up a storm if taken too far from the government tit. As Marc Reisner makes clear in *The Cadillac Desert*, much of the agriculture and ranching of the West exists by virtue of a complicated and very expensive sort of government welfare: big dams and water plans. The real intent of the circular (it urges people to write the state governor) seems to be to resist policies that favor long-range sustainability and the support of biodiversity, and to hold out for maximum resource extraction right now.

As far as I can see, the intelligent but so far toothless California "bioregional proposal" is simply a basis for further thinking and some degree of cooperation among agencies. The most original part is the call for the formation of "bioregional councils" that would have some stake in decision making. Who would be on the bioregional councils is not spelled out. Even closer to the roots, the memorandum that started all this furor suggests that "watershed councils" would be formed, which, being based on stream-by-stream communities, would be truly local bodies that could help design agreements working for the preservation of natural variety. Like, let's say, helping to preserve the spawning grounds for the wild salmon that still come (amazingly) into the lower Yuba River gravel wastelands. This would be an effort that would have to involve a number of groups and agencies, and it would have to include the blessing of the usually development-minded Yuba County Water Agency.

The term *bioregion* was adopted by the signers to the Memorandum on Biological Diversity as a technical term from the field of biogeography. It's not likely that they would have known that there were already groups of people around the United States and Canada who were talking in terms of bioregionally oriented societies. I doubt they would have heard about the first North American Bioregional Congress held in Kansas in the late eighties. They had no idea that for twenty years communitarian ecology-minded dwellers-in-the-land have been living in places they call "Ish" (Puget Sound and lower British Columbia) or "Columbiana" (upper Columbia River) or "Mesechabe" (lower Mississippi), or "Shasta" (northern California), and all of them have produced newsletters, taken field trips, organized gatherings, and at the same time participated in local politics.

That "bioregion" was an idea already in circulation was the bad, or good, luck of the biodiversity agreement people, depending on how you look at it. As it happens, the bioregional people are also finding "watershed councils" to be the building blocks of a long-range strategy for social and environmental sustainability.

A watershed is a marvelous thing to consider: this process of rain falling, streams flowing, and oceans evaporating causes every molecule of water on earth to make the complete trip once every two million years. The surface is carved into watersheds—a kind of familial branching, a chart of relationship, and a definition of place. The watershed is the first and last nation whose boundaries, though subtly shifting, are unarguable. Races of birds, subspecies of trees, and types of hats or rain gear often go by the watershed. For the watershed, cities and dams are ephemeral and of no more account than a boulder that falls in the river or a landslide that temporarily alters the channel. The water will always be there, and it will always find its way down. As constrained and polluted as the Los Angeles River is at the moment, it can also be said that in the larger picture that river is alive and well under the city streets, running in giant culverts. It may be amused by such diversions. But we who live in terms of centuries rather than millions of years must hold the watershed and its communities together, so our children might enjoy the clear water and fresh life of this landscape we have chosen. From the tiniest rivulet at the crest of a ridge to the main trunk of a river approaching the lowlands, the river is all one place and all one land.

The water cycle includes our springs and wells, our Sierra snowpack, our irrigation canals, our car wash, and the spring salmon run. It's the spring peeper in the pond and the acorn woodpecker chattering in a snag. The watershed is beyond the dichotomies of orderly/disorderly, for its forms are free, but somehow inevitable. The life that comes to flourish within it constitutes the first kind of community.

The agenda of a watershed council starts in a modest way: like saying, "Let's try and rehabilitate our river to the point that wild salmon can successfully spawn here again." In pursuit of this local agenda, a community might find itself combating clear-cut timber sales upstream, water-selling grabs downstream, Taiwanese drift-net practices out in the North Pacific, and a host of other national and international threats to the health of salmon.

If a wide range of people will join in on this effort— people from timber and tourism, settled ranchers and farmers, fly-fishing retirees, the businesses and the forest-dwelling new settlers—something might come of it. But if this joint agreement were to be implemented as a top-down prescription, it would go nowhere. Only a grass-roots engagement with long-term land issues can provide the political and social stability it will take to keep the biological richness of California's regions intact.

All public land ownership is ultimately written in sand. The boundaries and the management categories were created by Congress, and Congress can take them away. The only "jurisdiction" that will last in the world of nature is the watershed, and even that changes slightly over time. If public lands come under greater and greater pressure to be opened for exploitation and use in the twenty-first century, it will be the local people, the watershed people, who will prove to be the last and possibly most effective line of defense. Let us hope it never comes to that.

The mandate of the public land managers and the Fish and Wildlife people inevitably directs them to resource concerns. They are proposing to do what could be called "ecological bioregionalism." The other movement, coming out of the local communities, could be called "cultural bioregionalism." I would like to turn my attention now to cultural bioregionalism and to what practical promise these ideas hold for fin-de-millennium America.

Living in a place—the notion has been around for decades and has usually been dismissed as provincial, backward, dull, and possibly reactionary. But new dynamics are at work. The mobility that has characterized American life is coming to a close. As Americans begin to stay put, it may give us the first opening in over a century to give participatory democracy another try.

Daniel Kemmis, the mayor of Missoula, Montana, has written a fine little book called *Community and the Politics of Place* (Norman: University of Oklahoma Press, 1990). Mr. Kemmis points out that in the eighteenth century the word *republican* meant a politics of community engagement. Early republican thought was set against the federalist theories that would govern by balancing competing interests, devise sets of legalistic procedures, maintain checks and balances (leading to hearings held before putative experts) in place of direct discussion between adversarial parties.

Kemmis quotes Rousseau: "Keeping citizens apart has become the first maxim of modern politics." So what organizing principle will get citizens back together? There are many, and each in its way has its use. People have organized themselves by ethnic background, religion, race, class, employment, gender, language, and age. In a highly mobile society where few people stay put, thematic organizing is entirely understandable. But place, that oldest of organizing principles (next to kinship), is a novel development in the United States.

"What holds people together long enough to discover their power as citizens is their common inhabiting of a single place," Kemmis argues. Being so placed, people will volunteer for community projects, join school boards, and accept nominations and appointments. Good minds, which are often forced by company or agency policy to keep moving, will make notable contributions to the neighborhood if allowed to stay put. And since local elections deal with immediate issues, a lot more people will turn out to vote. There will be a return of civic life.

This will not be "nationalism" with all its danger, as long as sense of place is not entirely conflated with the idea of a nation. Bioregional concerns go beyond those of any ephemeral (and often brutal and dangerous) politically designated space. They give us the imagination of "citizenship" in a place called (for example) the great Central Valley, which has valley oaks and migratory waterfowl as well as humans among its members. A place (with a climate, with bugs), as Kemmis says, "develops practices, creates culture."

Another fruit of the enlarged sense of nature that systems ecology and bioregional thought have given us is the realization that cities and suburbs are all part of the system. Unlike the ecological bioregionalists, the cultural bioregionalists absolutely must include the cities in their thinking. The practice of urban bioregionalism ("green cities") has made a good start in San Francisco. One can learn and live deeply with regard to wild systems in any sort of neighborhood—from the urban to a big sugar-beet farm. The birds are migrating, the wild plants are looking for a way to slip in, the insects in any case live an untrammeled life, the raccoons are padding through the crosswalks at 2:00 A.M., and the nursery trees are trying to figure out who they are. These are exciting, convivial, and somewhat radical knowledges.

An economics of scale can be seen in the watershed/bioregion/city-state model. Imagine a Renaissance-style city-state facing out on the Pacific with its bioregional hinterland reaching to the headwaters of all the streams that flow through its bay. The San Francisco/valley rivers/Shasta headwaters bio-city-region! I take some ideas along these lines from Jane Jacobs's tantalizing book, *Cities and the Wealth of Nations* (New York: Random House, 1984), in which she argues that the city, not the nation-state, is the proper locus of an economy, and then that the city is always to be understood as being one with the hinterland.

Such a non-nationalistic idea of community, in which commitment to pure place is paramount, cannot be ethnic or racist. Here is perhaps the most delicious turn that comes out of thinking about politics from the standpoint of place: anyone of any race, language, religion, or origin is welcome, as long as they live well on the land. The great Central Valley region does not prefer English over Spanish or Japanese or Hmong. If it had any preferences at all, it might best like the languages it heard for thousands of years, such as Maidu or Miwok, simply because it's used to them. Mythically speaking, it will welcome whoever chooses to observe the etiquette, express the gratitude, grasp the tools, and learn the songs that it takes to live there.

This sort of future culture is available to whoever makes the choice, regardless of background. It need not require that a person drop his or her Buddhist, Jewish, Christian, animist, atheist, or Muslim beliefs but simply add to that faith or philosophy a sincere nod in the direction of the deep value of the natural world and the subjecthood of nonhuman beings. A culture of place will be created that will include the "United States," and go beyond that to an affirmation of the continent, the land itself, Turtle Island. We could be showing Southeast Asian and South American newcomers the patterns of the rivers, the distant hills, saying, "It is not only that you are now living in the United States. You are living in this great landscape. Please get to know these rivers and mountains, and be welcome here." Euro-Americans, Asian Americans, African Americans can—if they wish— become "born-again" natives of Turtle Island. In doing so we also might even (eventually) win some respect from our Native American predecessors, who are still here and still trying to teach us where we are.

Watershed consciousness and bioregionalism is not just environmentalism, not just a means toward resolution of social and economic problems, but a move toward resolving both nature and society with the practice of a profound citizenship in both the natural and the social worlds. If the ground can be our common ground, we can begin to talk to each other (human and nonhuman) once again.

California is gold-tan grasses, silver-gray tule fog,
olive-green redwood, blue-gray chaparral,
silver-hue serpentine hills.
Blinding white granite,
blue-black rock sea cliffs.
—Blue summer sky, chestnut brown slough water,
steep purple city streets—hot cream towns.
Many colors of the land, many colors of the skin.

[This essay was first given as a talk for the California Studies Center at Sacramento State College, as part of their conference entitled "Dancing at the Edge," on February 6, 1992. It was published in the San Francisco Examiner *on March 1 and 2, 1992, and was soon reprinted in a number of other periodicals and anthologies in the United States and England.]*

The Rediscovery of
Turtle Island

*For John Wesley Powell, watershed visionary,
and for Wallace Stegner*

I

We human beings of the developed societies have once more been expelled from a garden—the formal garden of Euro-American humanism and its assumptions of human superiority, priority, uniqueness, and dominance. We have been thrown back into that other garden with all the other animals and fungi and insects, where we can no longer be sure we are so privileged. The walls between "nature" and "culture" begin to crumble as we enter a posthuman era. Darwinian insights force occidental people, often unwillingly, to acknowledge their literal kinship with critters.

Ecological science investigates the interconnections of organisms and their constant transactions with energy and matter. Human societies come into being along with the rest of nature. There is no name yet for a humanistic scholarship that embraces the nonhuman. I suggest (in a spirit of pagan play) we call it "panhumanism."

Environmental activists, ecological scientists, and panhumanists are still in the process of reevaluating how to think about, how to create policy with, nature. The professional resource managers of the Forest Service and the Bureau of Land Management have been driven (partly by people of conscience within their own ranks) into rethinking their old utilitarian view of the vast lands in their charge. This is a time of lively confluence, as scientists, self-taught ecosystem experts from the communities, land management agency experts, and a new breed of ecologically aware loggers and ranchers (a few, but growing) are beginning to get together.

In the more rarefied world of ecological and social theory, the confluence is rockier. Nature writing, environmental history, and ecological philosophy have become subjects of study in the humanities. There are, however, still a few otherwise humane historians and philosophers who unreflectingly assume that the natural world is primarily a building-supply yard for human projects. That is what the Occident has said and thought for a couple thousand years.

Right now there are two sets of ideas circling about each other. One group, which we could call the "Savers," places value on extensive preservation of wilderness areas and argues for the importance of the original condition of nature.

This view has been tied to the idea that the mature condition of an ecosystem is a stable and diverse state technically called "climax." The other position holds that nature is constantly changing, that human agency has altered things to the point that there is no "natural condition" left, that there is no reason to value climax (or "fitness") over any other succession phase, and that human beings are not only part of nature but that they are also dominant over nature and should keep on using and changing it. They can be called the "Users." The Savers' view is attributed to the Sierra Club and other leading national organizations, to various "radical environmentalists," and to many environmental thinkers and writers. The Users' view, which has a few supporters in the biological sciences, has already become a favorite of the World Bank and those developers who are vexed by the problems associated with legislation that requires protection for creatures whose time and space are running out. It has been quickly seized on by the industry-sponsored pseudopopulist-flavored "Wise Use" movement.

Different as they are, both groups reflect the instrumentalist view of nature that has long been a mainstay of occidental thought. The Savers' idea of freezing some parts of nature into an icon of "pristine, uninhabited wilderness" is also to treat nature like a commodity, kept in a golden cage. Some preservationists have been insensitive to the plight of indigenous peoples whose home grounds were turned into protected wildlife preserves or parks, or to the plight of local workers and farmers who lose jobs as logging and grazing policies change.

The Users, in turn, are both pseudopopulist and multinational. On the local level they claim to speak for communities and workers (whose dilemma is real enough), but a little probing discloses industry funding. On the global scale their backers line up with huge forces of governments and corporations, with NAFTA and GATT, and raise the specter of further destruction of local communities. Their organizations are staffed by the sort of professionals whom Wendell Berry calls "hired itinerant vandals."

Postmodern theoreticians and critics have recently ventured into nature politics. Many of them have sided with the Users—they like to argue that nature is part of history, that human beings are part of nature, that there is little in the natural world that has not already been altered by human agency, that in any case our idea of "nature" is a projection of our social condition and that there is no sense in trying to preserve a theoretical wild. However, to say that the natural world is subject to continual change, that nature is shaped by history, or that our idea of reality is a self-serving illusion is not new. These positions still fail to come to grips with the question of how to deal with the pain and distress of real beings, plants and animals, as real as suffering humanity; and how to preserve natural variety. The need to protect worldwide biodiversity may be economically difficult and socially controversial, but there are strong scientific and practical arguments in support of it, and it is for many of us a profound ethical issue.

Hominids have obviously had some effect on the natural world, going back for half a million or more years. So we can totally drop the use of the word *pristine* in regard to nature as meaning "untouched by human agency." "Pristine" should now be understood as meaning "virtually" pristine. Almost any apparently untouched natural environment has in fact experienced some tiny degree of human impact. Historically there were huge preagricultural environments where the human impact, rather like deer or cougar activities, was normally almost invisible to any but a tracker's eye. The greatest single preagricultural human effect on wild nature, yet to be fully grasped, was the deliberate use of fire. In some cases human-caused fire seemed to mimic natural process, as with deliberate use of fire by native Californians. Alvar Núñez "Cabeza de Vaca," in his early-sixteenth-century walk across what is now Texas and the Southwest, found well-worn trails everywhere. But the fact still remains that there were great numbers of species, vast grasslands, fertile wetlands, and extensive forests in mosaics of all different stages in the preindustrial world. Barry Commoner has said that the greatest destruction of the world environment—by far—has taken place since 1950.

Furthermore, there is no "original condition" that once altered can never be redeemed. Original nature can be understood in terms of the myth of the "pool of Artemis"—the pool hidden in the forest that Artemis, goddess of wild things, visits to renew her virginity. The wild has—nay, *is*— a kind of hip, renewable virginity.

We are still laying the groundwork for a "culture of nature." The critique of the Judeo-Christian-Cartesian view of nature (by which complex of views all developed nations excuse themselves for their drastically destructive treatment of the landscape) is well under way. Some of us would hope to resume, reevaluate, re-create, and bring into line with complex science that old view that holds the whole phenomenal world to be our own being: multicentered, "alive" in its own manner, and effortlessly self-organizing in its own chaotic way. Elements of this view are found in a wide range of ancient vernacular philosophies, and it turns up in a variety of more sophisticated but still tentative forms in recent thought. It offers a third way, not caught up in the dualisms of body and mind, spirit and matter, or culture and nature. It is a noninstrumentalist view that extends intrinsic value to the nonhuman natural world.

Scouting parties are now following a skein of old tracks, aiming to cross and explore beyond the occidental (and postmodern) divide. I am going to lay out the case history of one of these probes. It's a potentially new story for the North American identity. It has already been in the making for more than thirty years. I call it "the rediscovery of Turtle Island."

II

In January 1969 I attended a gathering of Native American activists in southern California. Hundreds of people had come from all over the West. After sundown we went out to a gravelly wash that came down from the desert mountains. Drums were set up, a fire started, and for most of the night we sang the pantribal songs called "forty-nines." The night conversations circled around the idea of a native-inspired cultural and ecological renaissance for all of North America. I first heard this continent called "Turtle Island" there by a man who said his work was to be a messenger. He had his dark brown long hair tied in a Navajo men's knot, and he wore dusty khakis. He said that Turtle Island was the term that the people were coming to, a new name to help us build the future of North America. I asked him whom or where it came from. He said, "There are many creation myths with Turtle, East Coast and West Coast. But also you can just hear it."

I had recently returned to the West Coast from a ten-year residence in Japan. It was instantly illuminating to hear this continent renamed "Turtle Island." The realignments that conversation suggested were rich and complex. I was reminded that the indigenous people here have a long history of subtle and effective ways of working with their home grounds. They have had an exuberant variety of cultures and economies and some distinctive social forms (such as communal households) that were found throughout the hemisphere. They sometimes fought with each other, but usually with a deep sense of mutual respect. Within each of their various forms of religious life lay a powerful spiritual teaching on the matter of human and natural relationships, and for some individuals a practice of self-realization that came with trying to see through nonhuman eyes. The landscape was intimately known, and the very idea of community and kinship embraced and included the huge populations of wild beings. Much of the truth of Native American history and culture has been obscured by the self-serving histories that were written on behalf of the conquerors, the present dominant society.

This gathering took place one year before the first Earth

Day. As I reentered American life during the spring of 1969, I saw the use of the term "Turtle Island" spread through the fugitive Native American newsletters and other communications. I became aware that there was a notable groundswell of white people, too, who were seeing their life in the Western Hemisphere in a new way. Many whites figured that the best they could do on behalf of Turtle Island was to work for the environment, reinhabit the urban or rural margins, learn the landscape, and give support to Native Americans when asked. By 1970 I had moved with my family to the Sierra Nevada and was developing a forest homestead north of the South Yuba River. Many others entered the mountains and hills of the Pacific slope with virtually identical intentions, from the San Diego backcountry north into British Columbia. They had begun the reinhabitory move.

Through the early seventies I worked with my local forest community, but made regular trips to the cities, and was out on long swings around the country reading poems or leading workshops—many in urban areas. Our new sense of the Western Hemisphere permeated everything we did. So I called the book of poems I wrote from that period *Turtle Island* (New York: New Directions, 1974). The introduction says:

> *Turtle Island—the old-new name for the continent, based on many creation myths of the people who have been living here for millennia, and reapplied by some of them to "North America" in recent years. Also, an idea found worldwide, of the earth, or cosmos even, sustained by a great turtle or serpent-of-eternity.*

> *A name: that we may see ourselves more accurately on this continent of watersheds and life communities—plant zones, physiographic provinces, culture areas, following natural boundaries. The "U.S.A." and its states and counties are arbitrary and inaccurate impositions on what is really here.*
>
> *The poems speak of place, and the energy pathways that sustain life. Each living being is a swirl in the flow, a formal turbulence, a "song." The land, the planet itself, is also a living being—at another pace. Anglos, black people, Chicanos, and others beached up on these shores all share such views at the deepest levels of their old cultural traditions—African, Asian, or European. Hark again to those roots, to see our ancient solidarity, and then to the work of being together on Turtle Island.*

Following the publication of these poems, I began to hear back from a lot of people—many in Canada—who were remaking a North American life. Many other writers got into this sort of work each on his or her own—a brilliant and cranky bunch that included Jerry Rothenberg and his translation of Native American song and story into powerful little poem events; Peter Blue Cloud with his evocation of Coyote in a contemporary context; Dennis Tedlock, who offered a storyteller's representation of Zuni oral narrative in English; Ed Abbey, calling for a passionate commitment to the wild; Leslie Silko in her shivery novel *Ceremony*; Simon Ortiz in his early poems and stories—and many more.

A lot of this followed on the heels of the back-to-the-land

movement and the diaspora of longhairs and dropout graduate students to rural places in the early seventies. There are thousands of people from those days still making a culture: being teachers, plumbers, chair and cabinet makers, contractors and carpenters, poets in the schools, auto mechanics, geographic information computer consultants, registered foresters, professional storytellers, wildlife workers, river guides, mountain guides, architects, or organic gardeners. Many have simultaneously mastered grass-roots politics and the intricacies of public lands policies. Such people can be found tucked away in the cities, too.

The first wave of writers mentioned left some strong legacies: Rothenberg, Tedlock, and Dell Hymes gave us the field of ethnopoetics (the basis for truly appreciating multicultural literature); Leslie Silko and Simon Ortiz opened the way for a distinguished and diverse body of new American Indian writing; Ed Abbey's eco-warrior spirit led toward the emergence of the radical environmental group Earth First!, which (in splitting) later generated the Wild Lands Project. Some of my own writings contributed to the inclusion of Buddhist ethics and lumber industry work life in the mix, and writers as different as Wes Jackson, Wendell Berry, and Gary Paul Nabhan opened the way for a serious discussion of place, nature in place, and community. The Native American movement has become a serious player in the national debate, and the environmental movement has become (in some cases) big and controversial politics. Although the counterculture has faded and blended in, its fundamental concerns remain a serious part of the dialogue.

A key question is that of our ethical obligations to the nonhuman world. The very notion rattles the foundations of occidental thought. Native American religious beliefs, although not identical coast to coast, are overwhelmingly in support of a full and sensitive acknowledgment of the subjecthood—the intrinsic value—of nature. This in no way backs off from an unflinching awareness of the painful side of wild nature, of acknowledging how everything is being eaten alive. The twentieth-century syncretism of the "Turtle Island view" gathers ideas from Buddhism and Taoism and from the lively details of worldwide animism and paganism. There is no imposition of ideas of progress or order on the natural world—Buddhism teaches impermanence, suffering, compassion, and wisdom. Buddhist teachings go on to say that the true source of compassion and ethical behavior is paradoxically none other than one's own realization of the insubstantial and ephemeral nature of everything. Much of animism and paganism celebrates the actual, with its inevitable pain and death, and affirms the beauty of the process. Add contemporary ecosystem theory and environmental history to this, and you get a sense of what's at work.

Conservation biology, deep ecology, and other new disciplines are given a community constituency and real grounding by the bioregional movement. Bioregionalism calls for commitment to this continent *place by place*, in terms of biogeographical regions and watersheds. It calls us to see our country in terms of its landforms, plant life, weather patterns, and seasonal changes—its whole natural history before the net of political jurisdictions was cast over it. People are challenged to become "reinhabitory"—that is, to become people who are learning to live and think "as if" they were totally engaged with their place for the long future. This doesn't mean some return to a primitive lifestyle or utopian provincialism; it simply implies an engagement with community and a search for the sustainable sophisticated mix of economic practices that would enable people to live regionally and yet learn from and contribute to a planetary society. (Some of the best bioregional work is being done in cities, as people try to restore both human and ecological neighborhoods.) Such people are, regardless of national or ethnic backgrounds, in the process of becoming something deeper than "American (or Mexican or Canadian) citizens"—they are becoming natives of Turtle Island.

Now in the nineties the term "Turtle Island" continues, modestly, to extend its sway. There is a Turtle Island Office that moves around the country with its newsletter; it acts as a national information center for the many bioregional groups that every other year hold a "Turtle Island Congress." Participants come from Canada and Mexico as well as the United States. The use of the term is now standard in a number of Native American periodicals and circles. There is even a "Turtle Island String Quartet" based in San Francisco. In the winter of 1992 I practically convinced the director of the Centro de Estudios Norteamericanos at the Universidad de Alcalá in Madrid to change his department's name to "Estudios de la Isla de Tortuga." He much enjoyed the idea of the shift. We agreed: speak of the United States, and you are talking two centuries of basically English-speaking affairs; speak of "America" and you invoke five centuries of Euro-American schemes in the Western Hemisphere; speak of "Turtle Island" and a vast past, an open future, and all the life communities of plants, humans, and critters come into focus.

III

The Nisenan and Maidu, indigenous people who live on the east side of the Sacramento Valley and into the northern Sierra foothills, tell a creation story that goes something like this:

> *Coyote and Earthmaker were blowing around in the swirl of things. Coyote finally had enough of this aimlessness and said, "Earthmaker, find us a world!"*
>
> *Earthmaker tried to get out of it, tried to excuse himself, because he knew that a world can only mean trouble. But Coyote nagged him into trying. So leaning over the surface of the vast waters, Earthmaker called up Turtle. After a long time Turtle surfaced, and Earthmaker said, "Turtle, can you get me a bit of mud? Coyote wants a world."*
>
> *"A world," said Turtle. "Why bother? Oh, well." And down she dived. She went down and down and down, to the bottom of the sea. She took a great gob of mud, and started swimming toward the surface. As she spiraled and*

paddled upward, the streaming water washed the mud from the sides of her mouth, from the back of her mouth— and by the time she reached the surface (the trip took six years), nothing was left but one grain of dirt between the tips of her beak.

"That'll be enough!" said Earthmaker, taking it in his hands and giving it a pat like a tortilla. Suddenly Coyote and Earthmaker were standing on a piece of ground as big as a tarp. Then Earthmaker stamped his feet, and they were standing on a flat wide plain of mud. The ocean was gone. They stood on the land.

And then Coyote began to want trees and plants, and scenery, and the story goes on to tell how Coyote imagined landscapes that then came forth, and how he started naming the animals and plants as they appeared. "I'll call you skunk because you look like skunk." And the landscapes Coyote imagined are there today.

My children grew up with this as their first creation story. When they later heard the Bible story, they said, "That's a lot like Coyote and Earthmaker." But the Nisenan story gave them their own immediate landscape, complete with details, and the characters were animals from their own world.

Mythopoetic play can be part of what jump-starts long-range social change. But what about the short term? There are some immediate outcomes worth mentioning: a new era of community interaction with public lands has begun. In California a new set of ecosystem-based government/community joint-management discussions are beginning to take place. Some of the most vital environmental politics is being done by watershed or ecosystem-based groups. "Ecosystem management" by definition includes private landowners in the mix. In my corner of the northern Sierra, we are practicing being a "human-inhabited wildlife corridor"—an area that functions as a biological connector—and are coming to certain agreed-on practices that will enhance wildlife survival even as dozens of households continue to live here. Such neighborhood agreements would be one key to preserving wildlife diversity in most Third World countries.

Ultimately we can all lay claim to the term *native* and the songs and dances, the beads and feathers, and the profound responsibilities that go with it. We are all indigenous to this planet, this mosaic of wild gardens we are being called by nature and history to reinhabit in good spirit. Part of that responsibility is to choose a place. To restore the land one must live and work in a place. To work in a place is to work with others. People who work together in a place become a community, and a community, in time, grows a culture. To work on behalf of the wild is to restore culture.

[The University of California Humanities Research Institute sponsored a yearlong study called "Reinventing Nature" in 1992–93. Four conferences were held at four different campuses. It was an odd exercise: the occulted agenda seemed to ask that a critique of naive and sentimental environmentalism be done by postmodernist humanists and critics. The good-hearted humanists in fact had nothing against the environmentalists, and the conservation biology and bioregional enthusiasts (like me) told the narratives they knew best—our own goofy

breakaway revision of nature based on backpacking, zazen, Taoist parables, Coyote tales, half-understood cutting-edge science, Mahayana sutras, and field biology handbooks. This essay is based on a talk given at "Reinventing Nature/Recovering the Wild," the last conference in the series, held at the University of California at Davis, October 1993.]

Kitkitdizze: A Node in the Net

Jets heading west on the Denver-to-Sacramento run start losing altitude east of Reno, and the engines cool as they cross the snowy Sierra crest. They glide low over the west-tending mountain slopes, passing above the canyon of the north fork of the American River. If you look north out the window you can see the Yuba River country, and if it's really clear you can see the old "diggings"—large areas of white gravel laid bare by nineteenth-century gold mining. On the edge of one of those is a little hill where my family and I live. It's on a forested stretch between the South Yuba canyon and the two thousand treeless acres of old mining gravel, all on a forty-mile ridge that runs from the High Sierra to the valley floor near Marysville, California. You're looking out over the northern quarter of the greater Sierra ecosystem: a vast summer-dry hardwood-conifer forest, with drought-resistant shrubs and bushes in the canyons, clear-cuts, and burns.

In ten minutes the jet is skimming over the levees of the

Sacramento River and wheeling down the strip. It then takes two and a half hours to drive out of the valley and up to my place. The last three miles seem to take the longest—we like to joke that it's still the bumpiest road we've found, go where we will.

Back in the mid-sixties I was studying in Japan. Once, while I was on a visit to California, some friends suggested that I join them in buying mountain land. In those days land and gas were both still cheap. We drove into the ridge and canyon country, out to the end of a road. We pushed through manzanita thickets and strolled in open stretches of healthy ponderosa pine. Using a handheld compass, I found a couple of brass caps that mark corners. It was a new part of the Sierra for me. But I knew the assembly of plants—ponderosa pine, black oak, and associates—well enough to know what the rainfall and climate would be, and I knew I liked their company. There was a wild meadow full of native bunchgrass. No regular creek, but a slope with sedges that promised subsurface water. I told my friends to count me in. I put down the money for a twenty-five-acre share of the hundred acres and returned to Japan.

In 1969, back for good in California, we drove out to the land and made a family decision to put our life there. At that time there were virtually no neighbors, and the roads were even worse than they are now. No power lines, no phones, and twenty-five miles—across a canyon—to town. But we had the will and some of the skills as well. I had grown up on a small farm in the Northwest and had spent time in the forests and mountains since childhood. I had worked at carpentry and been a Forest Service seasonal worker, so mountain life (at three thousand feet) seemed doable. We weren't really "in the wilderness" but rather in a zone of ecological recovery. The Tahoe National Forest stretches for hundreds of square miles in the hills beyond us.

I had also been a logger on an Indian reservation in the ponderosa pine forests of eastern Oregon, where many trees were more than two hundred feet tall and five feet through. That land was drier and a bit higher, so the understory was different, but it grew the same adaptable cinnamon-colored pines. The trees down here topped out at about a hundred feet; they were getting toward being a mature stand, but a long way from old growth. I talked with a ninety-year-old neighbor who had been born in the area. He told me that when he was young he had run cattle over my way and had logged here and there, and that a big fire had gone through about 1920. I trimmed the stump on a black oak that had fallen and counted the rings: more than three hundred years. Lots of standing oaks that big around, so it was clear that the fires had not been total. Besides the pine stands (mixed with incense cedar, madrona, a few Douglas firs), our place was a mosaic of postfire manzanita fields with small pines coming through; stable climax manzanita; an eight-acre stand of pure black oak; and some areas of blue oak, gray pine, and grasses. Also lots of the low ground-cover bush called *kitkitdizze* in the language of the Wintun, a nearby valley people. It was clear from the very old and scattered stumps that this area had been selectively logged once. A neighbor with an increment borer figured that some trees had been cut about 1940. The surrounding lands and the place I was making my home flowed together with unmarked boundaries; to the eye and to the creatures, it was all one.

We had our hands full the first ten years just getting up walls and roofs, bathhouse, small barn, woodshed. A lot of it was done the old way: we dropped all the trees to be used in the frame of the house with a two-man falling saw, and peeled them with drawknives. Young women and men with long hair joined the work camp for comradeship, food, and spending money. (Two later became licensed architects; many of them stayed and are neighbors today) Light was from kerosene lamps; we heated with wood and cooked with wood and propane. Wood-burning ranges, wood-burning sauna stoves, treadle-operated sewing machines, and propane using Servel refrigerators from the fifties were the targets of highly selective shopping runs. Many other young settlers found their place in northern California in the early seventies, so eventually there was a whole reinhabitory culture living this way in what we like to call Shasta Nation.

I set up my library and wrote poems and essays by lantern light, then went out periodically, lecturing and teaching around the country. I thought of my home as a well-concealed base camp from which I raided university treasuries. We named our place Kitkitdizze after the aromatic little shrub.

The scattered neighbors and I started meeting once a month to talk about local affairs. We were all nature lovers, and everyone wanted to cause as little impact as possible. Those with well-watered sites with springs and meadows put in small gardens and planted fruit trees. I tried fruit trees, a chicken flock, a kitchen garden, and beehives. The bees went first. They were totally destroyed in one night by a black bear. The kitchen garden did fairly well until the run of dry winters that started in the eighties and may finally be over. And, of course, no matter how you fence a garden, deer find a way to get in. The chickens were constant targets of northern goshawks, red-tailed hawks, raccoons, feral dogs, and bobcats. A bobcat once killed twenty-five in one month. The fruit trees are still with us, especially the apples. They, of all the cultivars, have best made themselves at home. (The grosbeaks and finches always seem to beat us to the cherries.) · But in my heart I was never into gardening. I couldn't see myself as a logger again either, and it wasn't the place to grow Christmas trees. Except for cutting fallen oak and pine for firewood, felling an occasional pole for framing, and frequent clearing of the low limbs and underbrush well back from the homestead to reduce fire hazard, I hadn't done much with the forest. I wanted to go lightly, to get a deep sense of it, and thought it was enough to leave it wild, letting it be the wildlife habitat it is.

Living in a place like this is absolutely delicious. Coyote-howl fugues, owl exchanges in the treetops, the almost daily sighting of deer (and the rattle of antlers at rutting season), the frisson of seeing a poky rattlesnake, tracking critters in the snowfall, seeing cougar twice, running across humongous bear scats, sharing all this with the children are more than worth the inconveniences.

My original land partners were increasingly busy elsewhere. It took a number of years, but we bought our old partners out and ended up with the whole hundred acres.

That was sobering. Now Kitkitdizze was entirely in our hands. We were cash poor and land rich, and who needs more second-growth pine and manzanita? We needed to re-think our relation to this place with its busy—almost down-town—rush of plants and creatures. Do we leave it alone? Use it, but how? And what responsibility comes with it all?

Now it is two grown sons, two stepdaughters, three cars, two trucks, four buildings, one pond, two well pumps, close to a hundred chickens, seventeen fruit trees, two cats, about ninety cords of firewood, and three chainsaws later. I've learned a lot, but there still is plenty of dark and unknown territory. (There's one boundary to this land down in the chaparral—it borders the BLM—that I *still* haven't lo-cated.) Black bear leave pawprints on woodshed refrigera-tors, and bobcats, coyotes, and foxes are more in evidence than ever, sometimes strolling in broad daylight. Even the diggings, which were stripped of soil by giant nozzles wash-ing out the scattered gold, are colonized by ever-hardy manzanita and bonsai-looking pine. The first major envi-ronmental conflict in California was between Sacramento Valley farmers and the hydraulic gold miners of the Yuba. Judge Lorenzo Sawyer's decision of 1884 banned absolutely all release of mining debris into the watershed. That was the end of hydraulic mining here. We now know that the amount of material that was washed out of the Sierra into the valley and onto good farmlands was eight times the amount of dirt removed for the Panama Canal.

The kerosene lights have been replaced by a photovoltaic array powering a mixed AC/DC system. The phone com-pany put in an underground line to our whole area at its own expense. My wife Carole and I are now using computers, the writer's equivalent of a nice little chainsaw. (Chainsaws and computers increase both macho productivity and nerdy stress.) My part-time teaching job at the University of California, Davis, gives me an Internet account. We have entered the late twentieth century and are tapping into political and environmental information with a vengeance.

The whole Sierra is a mosaic of ownership—various national forests, Bureau of Land Management, Sierra Pacific Industries, state parks, and private holdings—but to the eye of a hawk it is one great sweep of rocks and woodlands. We, along with most of our neighbors, were involved in the forestry controversies of the last decade, particularly in regard to the long-range plans for the Tahoe National Forest. The county boosters still seem to take more pleasure in the romance of the gold era than in the subsequent processes of restoration. The Sierra foothills are still described as "Gold Country," the highway is called "49," there are businesses called "Nugget" and "Bonanza." I have nothing against gold—I wear it in my teeth and in my ear—but the real wealth here is the great Sierran forest. My neighbors and I have sat in on many hearings and had long and complicated discussions with silviculturalists, district rangers, and other experts from the Forest Service. All these public and private designations seem to come with various "rights." With just "rights" and no land ethic, our summer-dry forests could be irreversibly degraded into chaparral over the coming centuries. We were part of a nationwide campaign to reform forest practices. The upshot was a real and positive upheaval on a national scale in the U.S. Forest Service and the promise of ecosystem management, which if actualized as described would be splendid.

We next turned our focus to the nearby public lands managed by the BLM. It wasn't hard to see that these public lands were a key middle-elevation part of a passageway for deer and other wildlife from the high country to the valleys below. Our own holdings are part of that corridor. Then we were catapulted into a whole new game: the BLM area manager for central California became aware of our interest, drove up and walked the woods with us, talked with us, consulted with the community, and then said, "Let's cooperate in the long-range planning for these lands. We can share information." We agreed to work with him and launched a biological inventory, first with older volunteers and then with our own wild teenagers jumping in. We studied close to three thousand forested acres. We bushwhacked up and down the canyons to find out just what was there, in what combinations, in what quantity, in what diversity.

Some of it was tallied and mapped (my son Kai learned Geographical Information Systems techniques and put the data into a borrowed Sun Sparc workstation), and the rest of our observations were written up and put into bundles of notes on each small section. We had found some very large trees, located a California spotted owl pair, noted a little wetland with carnivorous sticky sundew, described a unique barren dome with serpentine endemics (plants that grow only in this special chemistry), identified large stands of vivid growing forest, and were struck by the tremendous buildup of fuel. The well-intended but ecologically ignorant fire-exclusion policies of the government agencies over the last century have made the forests of California an incredible tinderbox.

The droughty forests of California have been shaped for millennia by fire. A fire used to sweep through any given area, forest historians are now saying, roughly every twenty-five years, and in doing so kept the undergrowth down and left the big trees standing. The native people also deliberately started fires, so that the California forests of two hundred years ago, we are told, were structured of huge trees in parks that were fire-safe. Of course, there were always some manzanita fields and recovering burns, but overall there was far less fuel. To "leave it be wild" in its present state would be risking a fire that might set the land back to first-phase brush again. The tens of thousands of homes and ranches mixed among the wooded foothills down the whole Sierra front could burn.

The biological inventory resulted in the formation of the Yuba Watershed Institute, a nonprofit organization made up of local people, sponsoring projects and research on forestry, biodiversity, and economic sustainability with an eye to the larger region. One of the conclusions of the joint-management plan, unsurprisingly, was to try to reduce fuel load by every available means. We saw that a certain amount of smart selective logging would not be out of place, could help reduce fuel load, and might pay some of the cost of thinning and prescriptive burning. We named our lands, with the BLM's blessing, the 'Inimim Forest, from the Nisenan word for *pine*, in recognition of the first people here.

The work with fire, wildlife, and people extends through public and (willing) private parcels alike. Realizing that our area plays a critical biological role, we are trying to learn the ground rules by which humans might live together with animals in an "inhabited wildlife corridor." A project for netting and banding migrant songbirds during nest season (providing information for a Western Hemisphere database) is located on some Kitkitdizze brushlands, rather than public land, simply because it's an excellent location. It is managed by my wife, Carole, who is deeply touched by the spirit of the vibrant little birds she bands. Our cooperative efforts here can be seen as part of the rapidly changing outlook on land management in the West, which is talking public-private partnership in a big way. Joint-management agreements between local communities and other local and committed interests, and their neighboring blocks of public lands, are a new and potent possibility in the project of responsibly "recovering the commons" region by region. The need for ecological literacy, the sense of home watershed, and a better understanding of our stake in public lands are beginning to permeate the consciousness of the larger society.

Lessons learned in the landscape apply to our own lands, too. So this is what my family and I are borrowing from the watershed work as our own Three-Hundred-Year Kitkitdizze Plan: We'll do much more understory thinning and then a series of prescribed burns. Some patches will be left untouched by fire, to provide a control. We'll plant a few sugar pines, and incense cedars where they fit (ponderosa pines will mostly take care of themselves), burn the ground under some of the oaks to see what it does for the acorn crop, burn some bunchgrass patches to see if they produce better basketry materials (an idea from the Native basket-weaving revival in California). We'll leave a percentage of dead oak in the forest rather than take it all for firewood. In the time of our seventh-generation granddaughter there will be a large area of fire-safe pine stands that will provide the possibility of the occasional sale of an incredibly valuable huge, clear, old-growth sawlog.

We assume something of the same will be true on surrounding land. The wildlife will still pass through. And visitors from the highly crowded lowlands will come to walk, study, and reflect. A few people will be resident on this land, getting some of their income from forestry work. The rest may come from the information economy of three centuries hence. There might even be a civilization with a culture of cultivating wildness.

You can say that this is outrageously optimistic. It truly is. But the possibility of saving, restoring, and wisely (yes!) using the bounty of wild nature is still with us in North America. My home base, Kitkitdizze, is but one tiny node in an evolving net of bioregional homesteads and camps.

Beyond all this studying and managing and calculating, there's another level to knowing nature. We can go about learning the names of things and doing inventories of trees, bushes, and flowers, but nature as it flits by is not usually seen in a clear light. Our actual experience of many birds and much wildlife is chancy and quick. Wildlife is often simply a call, a cough in the dark, a shadow in the shrubs. You can watch a cougar on a wildlife video for hours, but the real cougar shows herself but once or twice in a lifetime. One must be tuned to hints and nuances.

After twenty years of walking right past it on my way to chores in the meadow, I actually paid attention to a certain gnarly canyon live oak one day. Or maybe it was ready to show itself to me. I felt its oldness, suchness, inwardness, oakness, as if it were my own. Such intimacy makes you totally at home in life and in yourself. But the years spent working around that oak in that meadow and not really noticing it were not wasted. Knowing names and habits, cutting some brush here, getting firewood there, watching for when the fall mushrooms bulge out are skills that are of themselves delightful and essential. And they also prepare one for suddenly meeting the oak.

[*1995*]